Foundations of Stochastic Differential Equations in Infinite Dimensional Spaces

KIYOSI ITÔ

Gakushuin University
Tokyo

Foundations of Stochastic Differential Equations in Infinite Dimensional Spaces

SOCIETY FOR INDUSTRIAL AND APPLIED MATHEMATICS
PHILADELPHIA, PENNSYLVANIA

10 9 8 7 6 5 4

Library of Congress Catalog Card Number: 84-50502
ISBN: 978-0-898711-93-6

 is a registered trademark.

Contents

Preface

This monograph is an elaboration of my lectures presented at the CBMS-NSF Regional Conference on Stochastic Differential Equations in Infinite Dimensional Spaces and Their Applications, held at Louisiana State University, Baton Rouge in 1983. Professor H. H. Kuo of Louisiana State University arranged for me to give a series of lectures at the conference, and to publish these lecture notes as part of the monograph series produced by the Conference Board of the Mathematical Sciences. I would like to express my sincere thanks to Professor Kuo for his kindness, and to the National Science Foundation and the Conference Board for their generous support.

I am much obliged to Louisiana State University and to the organizers of the conference, Professors Kuo and J. R. Doroh, for the superb arrangements, which enabled me to profit so much from the conference.

I am grateful also to Professor H. Föllmer and the Mathematical Institute, ETH, Zürich. A few months before the Regional Conference Professor Föllmer had kindly arranged for me to give a series of lectures at the Institute on the same subject, which enabled me to prepare the lectures for the conference.

Finally, I am grateful to Mrs. H. Shinohara for the painstaking job of typing my manuscript.

The infinite dimensional spaces where stochastic differential equations are formulated may be any function spaces. Here I have chosen the Schwartz spaces of distributions. I hope that the reader will appreciate that this choice makes it easier to connect general theory with concrete problems.

Kiyosi Itô
Gakushuin University, Tokyo
December 1983

Introduction

Throughout this monograph we assume that the base probability space is a *perfect probability measure space* $(\Omega, \mathscr{F}, P)$ and observe only the random variables that take values in *standard measurable spaces*. Let X be an $(E, \mathscr{E})$-valued random variable, namely a map $X: \Omega \to E$ measurable $\mathscr{D}(P)/\mathscr{E}$. The probability distribution of X,

$$P^X(B) = P(X^{-1}(B)), \qquad \mathscr{D}(P^X) = \{B : X^{-1}(B) \in \mathscr{D}(P)\}, \tag{1}$$

is a *Lebesgue probability measure* on $(E, \mathscr{E})$, the Lebesgue extension of a *Borel probability measure* on $(E, \mathscr{E})$. Under this assumption the *Kolmogorov extension theorem* holds for a sequence of random variables. See § 2.2 for details.

Let X be an $\mathbf{R}^d$-valued random variable. Then its probability distribution P^X is a Lebesgue probability measure on $(\mathbf{R}^d, \mathscr{B}(\mathbf{R}^d))$ which is characterized by its Fourier transform $\mathscr{F}P^X(z)$ (*characteristic function*), which in turn is equal to $E(e^{i(z, X)})$, $z \in \mathbf{R}^d$. The same fact holds for any random variable represented by a σ-concentrated P-measurable function that takes values in the dual space of a multi-Hilbertian space or more concretely in the Schwartz spaces $\mathscr{D}'$.

Some problems arise on such infinite dimensional random variables that never appear in the finite dimensional case. To fix the idea we observe a $(\mathscr{D}', \mathscr{B}_K(\mathscr{D}'))$-valued random variable where $\mathscr{D}' = \mathscr{D}'(\mathbf{R})$ and $\mathscr{B}_K(\mathscr{D}')$ is the Kolmogorov σ-algebra on $\mathscr{D}'$. $\mathscr{D}$ is a topological vector space with the Schwartz topology, denoted by $\tau = \tau(\mathscr{D})$. τ is determined by a directed family of Hilbertian seminorms weaker than τ (written $< \tau$; see § 1.5). An important fact is that for every $p < \tau$ we can find another Hilbertian seminorm $q < \tau$ such that p is weaker than q in the Hilbert–Schmidt sense, written $p <_{HS} q$, namely that $\sum p(e_n)^2 < \infty$ for every q-ONB $\{e_n\}$ in the Hilbertian seminormed space $(\mathscr{D}, q)$.

Let $X = \{X(f, \omega)\}$ be a $\mathscr{D}'$-*valued random variable* (abbr. $\mathscr{D}'$-variable). Then for every ω the map (the *sample functional*)

$$X_\omega : \mathscr{D} \to \mathbf{R}, \qquad f \mapsto X(f, \omega) \tag{2}$$

is linear and τ-continuous. Hence the map

$$X: \mathscr{D} \to L_0(\Omega) \equiv L_0(\Omega, P), \qquad f \mapsto X(f, \cdot) \tag{3}$$

turns out to be linear and τ-continuous.

Suppose that we are given a τ-continuous linear map $X: \mathscr{D} \to L_0(\Omega)$. This means that $\{X(f), f \in \mathscr{D}\}$ is a family of real random variables such that

$$X(a_1 f_1 + a_2 f_2) = a_1 X(f_1) + a_2 X(f_2) \quad \text{a.s.} \tag{4}$$

and

$$\lim_{f \underset{\tau}{\to} 0} E(|X(f)| \wedge 1) \to 0. \tag{5}$$

xi

But $X(f, \omega)$ is neither linear nor τ-continuous in f in general. Hence $\omega \mapsto x(\cdot, \omega)$ does not determine any $\mathscr{D}'$-variable. Nevertheless, the regularization theorem guarantees that there exists a unique $\mathscr{D}'$-variable $\tilde{X} = \{\tilde{X}(f, \omega)\}$ such that $\tilde{X}(f, \omega) = X(f, \omega)$ a.s. for every $f \in \mathscr{D}$. This random variable $\tilde{X}$ is called a *regular version* of X. (See §§ 2.3, 2.5 for details.)

A family of real random variables $X = \{X(f, \omega), f \in \mathscr{D}\}$ is called a τ-continuous *linear random functional* if $f \mapsto X(f)$ is a τ-continuous linear map $\mathscr{D} \to L_0(\Omega)$, whereas X is called a $\mathscr{D}'$-*variable* if for every ω the map $f \mapsto X(f, \omega)$ is a τ-continuous linear functional on $\mathscr{D}$. This distinction is similar to that between a stochastic process and its separable version; such a distinction is not necessary between a countable family of real random variables and its joint variable.

In many cases a linear random functional $(X(f), f \in \mathscr{D})$ satisfies

$$E(|X(f)|^2) \leqq cp(f)^2 \tag{6}$$

for some $p < \tau$. In this case the regular version $\tilde{X}$ of X satisfies

$$P\{\tilde{X} \in \mathscr{D}'_q\} = 1 \tag{7}$$

by the regularization theorem, where $p <_{HS} q < \tau$ and $\mathscr{D}'_q$ is the subspace of $\mathscr{D}'$ that consists of all q-continuous linear functionals on $\mathscr{D}$. $(\mathscr{D}'_q, q')$ ($q' =$ the dual norm of q) is a separable Hilbert space. It is often the case that because of some additional properties of X that enable us to use Kolmogorov's continuous version theorem we can obtain

$$P\{\tilde{X} \in C \cap \mathscr{D}'_q\} = 1, \tag{8}$$

where $C = C(\mathbf{R})$ is regarded as a subspace of $\mathscr{D}'$ in the obvious way.

Let $(\bar{\mathscr{D}}_p, \bar{p})$ and $(\bar{\mathscr{D}}_q, \bar{q})$ be the completions of $(\mathscr{D}, p)$ and $(\mathscr{D}, q)$ respectively, where $p <_{HS} q$. A linear random functional X on $\mathscr{D}$ satisfying (6) can be extended to a linear random functional $\bar{X}$ on $\bar{\mathscr{D}}_p$ satisfying

$$E(\bar{X}(f)^2) \leqq c\bar{p}(\bar{f})^2. \tag{9}$$

$\bar{X}$ is often denoted by the same notation X. The regular version $\tilde{X}$ can be extended to a $\bar{q}$-continuous linear functional $\bar{\tilde{X}}$ for almost every ω by virtue of (7). $\bar{\tilde{X}}$ is often denoted by the same notation $\tilde{X}$.

We must also distinguish processes of linear random functionals on $\mathscr{D}$ from $\mathscr{D}'$-valued stochastic processes, and similarly for stochastic differential equations. Usually it is easier to deal with the former ones, but once we obtain a result about the former ones, we can transform it into a result about the latter ones, using the regularization theorem.

Note. Here we use *positive definite* and *strictly positive definite* instead of nonnegative definite and positive definite respectively.

A *Borel probability* measure on a measurable space $(E, \mathscr{E})$ is a probability measure on E whose domain of definition is $\mathscr{E}$. The Lebesgue extension of a

Borel probability measure is called a *Lebesgue probability measure*. To avoid confusion, the Lebesgue measure in the classical sense is called the classical Lebesgue measure.

Throughout this monograph vector spaces, Hilbert spaces, L_p-spaces, and so forth are considered over the reals unless stated otherwise. References are listed at the end of the monograph.

CHAPTER 1

Multi-Hilbertian Spaces and Their Dual Spaces

1.1. Hilbertian seminorms. Let S be a vector space. A seminorm $p: S \to (0, \infty)$ is called a *Hilbertian* seminorm (abbr. H-seminorm) if p has the following property:

$$p(x+y)^2 + p(x-y)^2 = 2p(x)^2 + 2p(y)^2.$$

A bilinear functional $b: S \times S \to \mathbf{R}$ is called *positive definite* if $b(x, x) \geqq 0$. If furthermore $b(x, x) > 0$ for $x \neq 0$, b is called *strictly positive definite*. The correspondence between the H-seminorms $\{p\}$ and the positive definite bilinear functionals $\{b\}$ is 1–1 by

$$b(x, y) = \tfrac{1}{4}(p(x+y) - p(x-y)), \qquad p(x) = \sqrt{b(x, y)},$$

by virtue of Neumann's theorem [32]. Hence the $b(x, y)$ corresponding to $p(x)$ is denoted by $p(x, y)$ unless stated otherwise. $p(x, y)$ is strictly positive definite (and so $p(x, y)$ is an *inner product*) if and only if $p(x)$ is a norm (i.e. $p(x) > 0$ for $x \neq 0$). In this case (S, p) is a pre-Hilbert space.

A vector space S endowed with an H-seminorm p is called a *Hilbertian seminormed space* (S, p). (S, p) is a semimetric space with semimetric

$$d_p(x, y) = p(x - y).$$

The d_p-topology is often called the *p-topology*. If (S, p) has a countable dense subset, then (S, p) (or p) is called *separable*. Orthogonality, orthonormality and an orthonormal base (abbr. ONB) are defined on (S, p) in the same way as in Hilbert spaces.

Let (S, p) be a separable Hilbertian seminormed space. Then

$$\tilde{S}_p := S/N_p \qquad (N_p = \{x \in S : p(x) = 0\})$$

is a separable pre-Hilbert space with the Hilbertian norm $\tilde{p}$ induced by p in the obvious way. The completion of $(\tilde{S}_p, \tilde{p})$ is a separable Hilbert space, denoted by $(\bar{S}_p, \bar{p})$.

If p is a separable H-seminorm, then so is αp ($\alpha > 0$). If p_i, $i = 1, 2, \ldots, n$ are separable H-seminorms, then so is

$$p := \left(\sum_{i=1}^{n} p_i^2 \right)^{1/2}.$$

This H-seminorm has the property that the p-topology on S is the weakest of all the topologies that are stronger than every p_i-topology ($i = 1, 2, \ldots, n$). In view of this property we denote p by

$$\bigvee_{i=1}^{n} p_i.$$

DEFINITION 1.1.1. Let p and q be separable H-seminorms on S.

$$(p:q) := \sup\{p(x):q(x) \leqq 1\},$$

$$(p:q)_{HS} := \left(\sum_n p(e_n)^2\right)^{1/2} \quad (\{e_n\}: \text{an ONB on } (S,q))$$

if $(p:q) < \infty$. Otherwise, we set $(p:q)_{HS} = \infty$.

Remark 1.1.1. $(p:q)_{HS}$ is well defined independently of the choice of $\{e_n\}$. (See Remark 1.1.2 for the proof.)

DEFINITION 1.1.2. Let p and q be separable H-seminorms on S. p is said to be *bounded* by q, written $p \prec q$, if $(p:q) < \infty$. p is said to be *Hilbert–Schmidt bounded* (abbr. HS bounded) by q, written $p \prec_{HS} q$, if $(p:q)_{HS} < \infty$.

THEOREM 1.1.1. *Let $p, q, r, \ldots$ be separable H-seminorms on a vector space S. Then we have the following:*

(i)
$$(p:q) \leqq (p:q)_{HS},$$

and so
$$p \prec_{HS} q \Rightarrow p \prec q.$$

(ii)
$$(p:r) \leqq (p:q)(q:r),$$

and so
$$p \prec q \prec r \Rightarrow p \prec r.$$

(iii)
$$(p:r)_{HS} \leqq (p:q)(q:r)_{HS},$$
$$(p:r)_{HS} \leqq (p:q)_{HS}(q:r),$$

and so
$$p \prec q \prec_{HS} r \Rightarrow p \prec_{HS} r,$$
$$p \prec_{HS} q \prec r \Rightarrow p \prec_{HS} r.$$

(iv)
$$p_i \prec_{HS} r \ (i=1,2,\ldots,n) \Rightarrow \bigvee_{i=1}^{n} \alpha_i p_i \prec_{HS} r \ (\alpha_i > 0).$$

Proof [9]. (i) Suppose that $q(x) > 0$. Then we can use the Schmidt orthogonalization to find an ONB $\{e_1, e_2, \ldots\}$ on (S, q) such that $e_1 = x/q(x)$. Then

$$(p:q)_{HS}^2 = \sum_n p(e_n)^2 \geqq p(e_1)^2 = p(x)^2/q(x)^2.$$

Hence $(p:q)_{HS} \geqq (p:q)$.

(ii) and (iv) are easy.

(iii) It suffices to discuss the case where $(p:q)$, $(q:r)$ and $(p:r)$ are finite. Then

$$H_1 = (\bar{S}_p, \bar{p}), \quad H_2 = (\bar{S}_q, \bar{q}), \quad H_3 = (\bar{S}_r, \bar{r})$$

are separable Hilbert spaces, and the identity operator $I: S \to S$ induces a bounded linear operator $I_{pq}: H_2 \to H_1$. From Definition 1.1.1 we can easily deduce that

$$(p:q) = \|I_{pq}\| \quad \text{and} \quad (p:q)_{HS} = \|I_{pq}\|_{HS} \tag{1}$$

where $\|A\|$ and $\|A\|_{HS}$ are the norm and the Hilbert–Schmidt norm (abbr. HS-norm) of the operator A. Similarly we observe $I_{qr}: H_3 \to H_2$ and $I_{pr}: H_3 \to H_1$. Then

$$I_{pr} = I_{pq} \cdot I_{qr}.$$

Let $\{e_n\}$ be an ONB on H_3. Then

$$\bar{p}(I_{pr}e_n)^2 = \bar{p}(I_{pq}I_{qr}e_n)^2 \leqq \|I_{pq}\|^2 \, \bar{q}(I_{qr}e_n)^2.$$

Summing for $n = 1, 2, \ldots$, we obtain

$$\|I_{pr}\|^2_{HS} \leqq \|I_{pq}\|^2 \|I_{qr}\|^2_{HS}, \tag{2}$$

i.e.,

$$(p:r)_{HS} \leqq (p:q)(q:r)_{HS}.$$

To prove that

$$(p:r)_{HS} \leqq (p:q)_{HS}(q:r),$$

we will first observe that the conjugate operator $I^*_{pq}: H_1 \to H_2$ has the same norm and the same HS-norm as I_{pq}. Let $\{e_n\}$ and $\{d_n\}$ be ONB's on H_1 and H_2 respectively. Then

$$\sum_n \bar{q}(I^*_{pq}e_n)^2 = \sum_n \sum_m \bar{q}(I^*_{pq}e_n, d_m)^2 = \sum_n \sum_m \bar{p}(e_n, I_{pq}d_m) = \sum_m \bar{p}(I_{pq}d_m)^2, \tag{3}$$

i.e.,

$$\|I^*_{pq}\|_{HS} = \|I_{pq}\|_{HS}. \tag{4}$$

It is easy to see that $\|I^*_{pq}\| = \|I_{pq}\|$, and similarly for I_{qr} and I_{pr}. Since $I_{pr} = I_{pq}I_{qr}$, we have

$$I^*_{pr} = I^*_{qr}I^*_{pq}.$$

This implies that

$$\|I^*_{pr}\|_{HS} \leqq \|I^*_{qr}\| \|I^*_{pq}\|_{HS}$$

by the argument used in proving (2). But this, combined with (4), implies that

$$\|I_{pr}\|_{HS} \leqq \|I_{qr}\| \|I_{pq}\|_{HS},$$

i.e., $(p:r)_{HS} \leqq (p:q)_{HS}(q:r)$. $\quad\square$

Remark 1.1.2. The equality (3) observed above implies that $\sum_m \bar{p}(I_{pq}d_m)^2$ does not change even if $\{d_m\}$ is replaced by another ONB $\{d'_m\}$ on $H_2 = (S_q, \bar{q})$. This implies the fact mentioned in Remark 1.1.1.

1.2. Multi-Hilbertian spaces and dual multi-Hilbertian spaces. Let S be a vector space. A topology τ on S is called *multi-Hilbertian* if there exists a family of separable H-seminorms $\mathbf{p}$ such that the sets

$$\{y \in S : p_i(y-x) < \varepsilon_i, \, i = 1, 2, \ldots, n\} \qquad (n \in \mathbf{N}, \, p_i \in \mathbf{p}, \, \varepsilon_i > 0)$$

form a complete system of τ-neighborhoods of x for every $x \in S$. It is obvious that a vector space with a multi-Hilbertian topology is a special topological vector space, called a *multi-Hilbertian space*. The multi-Hilbertian topology determined by $\mathbf{p}$ is denoted by $\tau(\mathbf{p})$.

A family of separable H-seminorms $\mathbf{p}$ is *directed* if for every $p, q \in \mathbf{p}$ there exists $r \in \mathbf{p}$ such that $p, q < r$. If $\mathbf{p}$ is directed, then the following sets form a complete system of $\tau(\mathbf{p})$-neighborhoods of x:

$$\{y \in S : p(y-x) < \varepsilon\}, \qquad p \in \mathbf{p}, \quad \varepsilon > 0.$$

For every family $\mathbf{p}$ the family $\mathbf{q}$ consisting of all H-seminorms of the form

$$q = \bigvee_{i=1}^{n} p_i \qquad (n \in \mathbf{N}, \, p_i \in \mathbf{p})$$

determines the same topology as $\mathbf{p}$. Hence every multi-Hilbertian topology is determined by a directed family of separable H-seminorms. If $\mathbf{p}$ is *separating*, namely if for every $x \neq 0$ there exists $p \in \mathbf{p}$ such that $p(x) > 0$, then $\tau(\mathbf{p})$ is Hausdorff.

If $\mathbf{p}$ is a countable family, then $\tau(\mathbf{p})$ and $(S, \tau(\mathbf{p}))$ are called *countably Hilbertian*. If $\mathbf{p} = \{p_n, n = 1, 2, \ldots\}$, then $\tau(\mathbf{p})$ is determined by a countable directed family:

$$q_n = \bigvee_{i=1}^{n} p_i, \qquad n = 1, 2, \ldots .$$

In the special case where $\mathbf{p}$ consists of a single norm $(S, \tau(\mathbf{p}))$ is a separable pre-Hilbert space.

The following theorem is an immediate consequence of the property of a seminorm $r : r(\alpha x) = |\alpha| \, r(x)$.

THEOREM 1.2.1. *Let (S, τ) be a multi-Hilbertian space where $\tau = \tau(\mathbf{p})$ ($\mathbf{p}$: directed).*

(i) *If q is a separable Hilbertian seminorm on S such that $q < \mathbf{p}$, i.e., $\tau(q)$ is weaker than $\tau(\mathbf{p})$, then $q < p$ for some $p \in \mathbf{p}$.*

(ii) *If f is a continuous linear functional on (S, τ), then f is p-continuous for some $p \in \mathbf{p}$, i.e.*

$$|f(x)| \leqq cp(x) \quad \text{for some } c > 0.$$

Let τ be a multi-Hilbertian topology on S and let q be a separable H-seminorm on S. If the q-topology is weaker than τ, we write $q < \tau$. It is obvious that τ is determined by the family of all q's such that $q < \tau$. Let τ_1 and τ_2 be two multi-Hilbertian topologies on S. It is easy to see that τ_1 is *weaker* than τ_2, written $\tau_1 < \tau_2$, if and only if for every $p < \tau_1$ there exists $q < \tau_2$ such

that $p < q$. τ_1 is defined to be *Hilbert–Schmidt weaker* (abbr. *HS* weaker) than τ_2, written $\tau_1 <_{HS} \tau_2$, if for every $p < \tau_1$ there exists $q < \tau_2$ such that $p <_{HS} q$.
If $\tau_1 = \tau(\mathbf{p})$ and $\tau_2 = \tau(\mathbf{q})$ ($\mathbf{p}, \mathbf{q}$: directed), then

$$\tau_1 < \tau_2 \Leftrightarrow (\forall p \in \mathbf{p})(\exists q \in \mathbf{q})p < q,$$

$$\tau_1 <_{HS} \tau_2 \Leftrightarrow (\forall p \in \mathbf{p})(\exists q \in \mathbf{q})p <_{HS} q.$$

Let τ be a multi-Hilbertian topology on S. The multi-Hilbertian topology determined by

$$\{p : p <_{HS} q \text{ for some } q < \tau\}$$

is called the *Kolmogorov I-topology* of τ, denoted by $I(\tau)$ [17]. $I(\tau)$ is the strongest of all multi-Hilbertian topologies *HS* weaker than τ. In the special case where τ is determined by a single separable H-seminorm, then $I(\tau)$ is called the *Sazonov topology* [26] or the *Gross topology* of τ. If $I(\tau) = \tau$, then τ is called *nuclear* and (S, τ) is called a *nuclear space*.

The continuous linear functionals on a multi-Hilbertian space $S = (S, \tau)$ form a vector space with the usual linear operations, denoted by S'. Such a space S' is called a *dual multi-Hilbertian space*. If τ is countably Hilbertian, then S' is called a *dual countably Hilbertian space*.

Since $f(x)$, viewed as a functional of $f \in S'$ and $x \in S$, is bilinear, we often denote $f(x)$ by $\langle f, x \rangle$ or by $\langle x, f \rangle$. The *weak topology* τ_w on S' is given by the family of seminorms

$$w_x(f) = |f(x)|, \qquad x \in S,$$

and the *strong topology* τ_s on S' is given by the family of seminorms

$$s_B(f) = \sup \{|f(x)|, x \in B\}, \qquad B : \text{bounded},$$

where B is called *bounded*, if for every τ-neighborhood U of $0 \in S$ we can find $n = n(B, U)$ such that $nU \supset B$. Each of these topologies is a uniform Hausdorff topology and makes S' a topological vector space.

Let $S = (S, \tau)$ be a multi-Hilbertian space where $\tau = \tau(\mathbf{p})$ ($\mathbf{p}$: directed). For $p < \tau$ we define $S'_p \equiv (S, p)'$ to be the p-continuous linear functionals on S. S'_p is a separable Hilbert space with the dual norm p'

$$p'(f) = \sup \{|f(x)| : p(x) \leq 1\}.$$

$S'_p = (S'_p, p')$ is isomorphic to the separable Hilbert space $\bar{S}_p = (\bar{S}_p, \bar{p})$ introduced in the last section. Using Theorem 1.2.1 (ii), we obtain the following:

THEOREM 1.2.2.

$$S' = \bigcup_{p \in \mathbf{p}} S'_p = \bigcup_{q < \tau(\mathbf{p})} S'_q.$$

If $p \leq cq$, then $q' \leq cp'$. Also if B is bounded in S, then for every p there exists $c > 0$ such that

$$B \subset \{x \in S : p(x) < c\},$$

and so

$$s_B(f) \leqq cp'(f), \qquad f \in S'_p.$$

Hence we obtain the following:

THEOREM 1.2.3. (i) *If $p < q$, then $S'_p \subset S'_q$, and the inclusion map from (S'_p, p') into (S'_q, q') is continuous.*

(ii) *For every p, $S'_p \subset S'$, and the inclusion map from (S'_p, p') into (S', τ_s) is continuous.*

The σ-algebra on S' generated by the strongly (resp. weakly) open sets is called the *topological σ-algebra* with respect to the strong (resp. weak) topology denoted by $\mathscr{B}_s(S')$ (resp. $\mathscr{B}_w(S')$). Also the σ-algebra on S' generated by the half-spaces:

$$\{f \in S' : f(x) \leqq c\}, \qquad x \in S, \quad c \in \mathbf{R}$$

is called the *Kolmogorov σ-algebra $\mathscr{B}_K(S')$*. We have

$$\mathscr{B}_s(S') \supset \mathscr{B}_w(S') \supset \mathscr{B}_K(S')$$

in general. However, if $\mathbf{p}$ is countable, then these σ-algebras coincide with each others; in particular,

$$\mathscr{B}_s(S'_p) = \mathscr{B}_w(S'_p) = \mathscr{B}_K(S'_p).$$

In our present discussion we are mainly concerned with Kolmogorov σ-algebras.

THEOREM 1.2.4.

$$\mathscr{B}_K(S'_p) = \mathscr{B}_K(S') \cap S'_p \qquad (\equiv \{B \cap S'_p : B \in \mathscr{B}_K(S')\}).$$

1.3. $\mathscr{S}$ **and** $\mathscr{S}'$. Let $\mathscr{S} = \mathscr{S}(\mathbf{R})$ be the space of all rapidly decreasing functions [27]. The Schwartz topology τ in $\mathscr{S}$ is given by the norms

$$|\varphi|_{m,n} = \sup_t |t^m \varphi^{(n)}(t)|, \qquad m, n = 0, 1, 2, 3, \ldots.$$

These norms are not H-norms. But τ is also given by a family of H-norms, which will be introduced below.

Let $H_n(t)$, $n = 0, 1, 2, \ldots$ be the Hermite polynomials and set

$$h_n(t) = (2^n n! \sqrt{\pi})^{-1/2} e^{-t^2/2} H_n(t).$$

Then $\{h_n\}$ is an ONB in the Hilbert space $L_2 = L_2(\mathbf{R})$, on which the norm and the inner product are denoted by $\|\cdot\|$ and $(\cdot, \cdot)$ respectively. We define $\|f\|_p \in [0, \infty]$ by

$$\|f\|_p^2 = \sum_{n=0}^{\infty} (2n+1)^{2p} (f, h_n)^2, \qquad f \in L_2, \quad p \in \mathbf{R}.$$

$\|f\|_p$ increases with p. $\|f\|_p < \infty$ for $p \leqq 0$. $\|f\|_p < \infty$ for $f \in \mathscr{S}$ and $p \in \mathbf{R}$, and $\|f\|_p$ is an H-norm on $\mathscr{S}$ for every $p \in \mathbf{R}$. The corresponding inner product is given by

$$(f, g)_p = \sum_{n=0}^{\infty} (2n+1)^{2p}(f, h_n)(g, h_n), \qquad f, g \in \mathcal{S}.$$

$(\mathcal{S}, \|\cdot\|_p)$ is a pre-Hilbert space and

$$h_n^p = (2n+1)^{-p} h_n, \qquad n = 0, 1, 2, \ldots$$

form an ONB in $(\mathcal{S}, \|\cdot\|_p)$. The completion of $(\mathcal{S}, \|\cdot\|_p)$ is a separable Hilbert space, denoted by $(\mathcal{S}_p, \|\cdot\|_p)$. $\{h_n^p\}_n$ is also an ONB in $(\mathcal{S}_p, \|\cdot\|_p)$. $\mathcal{S}_p$ $(p \geqq 0)$ is realized as a subspace of L_2 as follows:

$$\mathcal{S}_p = \{f \in L_2 : \|f\|_p < \infty\}.$$

The Schwartz topology τ on $\mathcal{S}$ coincides with the countably Hilbertian topology determined by $\|\cdot\|_p$, $p = 1, 2, 3, \ldots$. It is obvious that

$$\|\cdot\|_p < \tau, \qquad p \in \mathbf{R}.$$

Since

$$\sum_n \|h_n^{p+1}\|_p^2 = \sum_n (2n+1)^{-2} < \infty,$$

we obtain

$$\|\cdot\|_p <_{HS} \|\cdot\|_{p+1}, \qquad p \in \mathbf{R},$$

which ensures that $I(\tau) = \tau$. Hence τ is nuclear.

Let $\mathcal{S}'$ be the dual of $(\mathcal{S}, \tau)$, the continuous linear functionals on $(\mathcal{S}, \tau)$. An element of $\mathcal{S}'$ is called a *tempered distribution*. Let $\|\alpha\|_p'$ $(p \in \mathbf{R})$ be the dual norm of $\alpha \in \mathcal{S}'$, namely

$$\|\alpha\|_p' = \sup \{|\alpha(f)| : \|f\|_p \leqq 1\}.$$

Then $\|\cdot\|_p'$ is an H-norm on

$$\mathcal{S}_p' = \{\alpha \in \mathcal{S}' : \|\alpha\|_p' < \infty\},$$

and $(\mathcal{S}_p', \|\cdot\|_p')$ is a separable Hilbert space, which is isomorphic to $(\mathcal{S}_p, \|\cdot\|_p)$ by the map

$$\theta_p : \mathcal{S}_p \to \mathcal{S}_p', \qquad f \mapsto (\theta_p f)(\cdot) \equiv (f, \cdot)_p.$$

A measurable function f is called *slowly increasing*, written $f \in \mathcal{F}$, if $f(t)(1+t^2)^{-n}$ is square integrable for some $n = 0, 1, 2, \ldots$. With a slowly increasing function f we associated a *tempered distribution*

$$\alpha_f(g) = \int_{\mathbf{R}} f(t)g(t)\, dt = (f, g), \qquad g \in \mathcal{S}.$$

Since the map $f \mapsto \alpha_f$ is injective from $\mathcal{F}$ into $\mathcal{S}'$, we usually identify f with α_f and embed $\mathcal{F}$ into $\mathcal{S}'$. Hereafter we assume that such an identification is made, so that $\mathcal{F}$ is a subset of $\mathcal{S}'$. Since $L_2 \subset \mathcal{F}$, we obtain

$$\mathcal{S} \subset \mathcal{S}_p \subset \mathcal{S}_0 = \mathcal{S}_0' = L_2 \subset \mathcal{S}_p' \subset \mathcal{S}', \qquad p \geqq 0.$$

If $f \in \mathscr{S}$, then $f\ (=\alpha_f) \in \mathscr{S}'_0$, and furthermore $f \in \mathscr{S}'_{-p}$, $p \geqq 0$, because

$$|(f, g)| \leqq \|f\|_p \, \|g\|_{-p}, \qquad g \in \mathscr{S}, \quad p \geqq 0$$

follows from the definition of $\|\cdot\|_p$. This implies that $\mathscr{S} \subset \mathscr{S}'_{-p}$, $p \geqq 0$. We will prove that

$$(\mathscr{S}_{-p}, \|\cdot\|_{-p}) = (\mathscr{S}_p, \|\cdot\|_p), \qquad p \geqq 0.$$

To prove this, it suffices to check that $\{h_n^p\}_n$ ($\subset \mathscr{S} \subset \mathscr{S}'_{-p} \cap \mathscr{S}_p$) is an ONB in $(\mathscr{S}'_{-p}, \|\cdot\|_{-p})$, because $\{h_n^p\}_n$ is an ONB in $(\mathscr{S}_p, \|\cdot\|_p)$. Since θ_{-p} is an isomorphism from $(\mathscr{S}_{-p}, \|\cdot\|_{-p})$ to $(\mathscr{S}'_{-p}, \|\cdot\|'_{-p})$, $\{\theta_{-p} h_n^{-p}\}_n$ is an ONB in $(\mathscr{S}'_{-p}, \|\cdot\|'_{-p})$. Using the definitions of $(\cdot, \cdot)_p$ and h_n^p, we obtain

$$(\theta_{-p} h_n^{-p})(f) = (h_n^{-p}, f)_{-p} = (h_n^p, f), \qquad f \in \mathscr{S},$$

so $\theta_{-p} h_n^{-p} = h_n^p$. Hence $\{h_n^p\}_n$ is an ONB in $(\mathscr{S}'_{-p}, \|\cdot\|'_{-p})$.

Since

$$D h_n = (2n+1) h_n, \qquad D = t^2 - \frac{d^2}{dt^2},$$

we can express $\|\cdot\|_p$ as follows:

$$\|\varphi\|_p = \|D^p \varphi\|, \qquad \varphi \in \mathscr{S}.$$

Since $(\mathscr{S}_p, \|\cdot\|_p)$ is the completion of $(\mathscr{S}, \|\cdot\|_p)$, we can use the Sobolev lemma to prove that

$$\mathscr{S} = \bigcap_{p=0}^{\infty} \mathscr{S}_p = \bigcap_{p=0}^{\infty} \mathscr{S}'_{-p}.$$

Similarly we can discuss $\mathscr{S}(\mathbf{R}^d)$ and $\mathscr{S}'(\mathbf{R}^d)$, where we use

$$h_{n_1 n_2 \cdots n_d}(t_1, t_2, \ldots, t_d) = \prod_{i=1}^{d} h_{n_i}(t_i)$$

instead of h_n.

Remark 1.3.1. τ is also determined by a countable family of Hilbertian norms:

$$\|\|f\|\|_p^2 = \sum_{m=0}^{p} \sum_{n=0}^{p} \int_{\mathbf{R}} (t^m f^{(n)}(t))^2 \, dt, \qquad p = 1, 2, 3, \ldots,$$

where

$$\|\|\cdot\|\|_p <_{HS} \|\|\cdot\|\|_{p+r} \quad \text{for some } r = r(p).$$

To prove this we expand f in terms of $\{h_n\}$ and use the well-known properties of $\{H_n\}$ to check that

$$\|\|\cdot\|\|_p < \|\cdot\|_{p+k} \quad \text{and} \quad \|\cdot\|_p < \|\|\cdot\|\|_{p+k}$$

for some $k = k(p)$.

1.4. The spaces $\mathscr{D}$ and $\mathscr{D}'$ on compact intervals. Let $\mathscr{D}$ be the vector space consisting of all functions on a compact interval $I = [a, b]$ that can be extended to C^∞ functions on $\mathbf{R}$ vanishing outside of I. For simplicity of notation we assume that $I = [0, \pi]$. We denote the norm and the inner product in $L_2 = L_2[0, \pi]$ by $\|\cdot\|$ and $(\cdot, \cdot)$ respectively. $\mathscr{D}$ is a dense vector subspace of L_2. We consider the following symmetric operator:

$$D = -\frac{d^2}{dt^2} : \mathscr{D} \to L_2.$$

We denote the closed extension of D by the same notation D.

The functions

$$s_n(t) := \sqrt{2/\pi} \sin nt, \qquad n = 1, 2, \dots$$

form an ONB in L_2. Since

$$Ds_n = n^2 s_n, \qquad n = 1, 2, \dots,$$

D is a strictly positive definite self-adjoint operator with spectral decomposition.

$$D = \sum_{n=1}^{\infty} n^2 E_n, \qquad E_n f = (f, s_n) s_n.$$

Hence D^p is well defined for $p \in \mathbf{R}$. $\mathscr{D}$ is contained in the domain of D^p for every $p \in \mathbf{R}$.

The norm

$$\|f\|_p := \|D^p f\|$$

defines an H-norm in $\mathscr{D}$ for every $p \in \mathbf{R}$. This norm is expressed in terms of $\{s_n\}$ as follows:

$$\|f\|_p^2 = \sum_{n=1}^{\infty} n^{4p} (f, s_n)^2.$$

The operator $K := D^{-1/2}$ is a strictly positive definite HS-operator in L_2 and

$$\|K\|_{HS}^2 = \sum_{n=1}^{\infty} n^{-2} < \infty.$$

The Schwartz topology τ on $\mathscr{D}$ [27] determined by the norms

$$|f|_p = \max_{n \leq p} \sup_{t \in I} |f^{(n)}(t)|, \qquad p = 1, 2, 3, \dots$$

is also determined by the above-mentioned H-norms $\|\cdot\|_p$, $p = 1, 2, 3, \dots$. Hence τ is countably Hilbertian. Since $\|\cdot\|_p$ increases with $p \in \mathbf{R}$, it is obvious that

$$\|\cdot\|_p \prec \tau, \qquad p \in \mathbf{R}.$$

We will prove that τ is nuclear (i.e. $I(\tau) = \tau$), so $(\mathscr{D}, \tau)$ is a nuclear space. To prove this, it suffices to check that

$$\|\cdot\|_p \prec_{HS} \|\cdot\|_{p+1/2}. \tag{1}$$

Using the Schmidt orthogonalization, we obtain an ONB $\{e_n\}$ in $(\mathscr{D}, \|\cdot\|)$; note that s_n does not belong to $\mathscr{D}$. Then $\{D^{-(p+1/2)}e_n, \ n = 1, 2, \ldots\}$ is an ONB in $(\mathscr{D}, \|\cdot\|_{p+1/2})$. Noting that $K = D^{-1/2}$, we obtain

$$\sum_n \|D^{-(p+1/2)}e_n\|_p^2 = \sum_n \|D^{-p}Ke_n\|_p^2 = \sum_n \|Ke_n\|^2 = \|K\|_{HS}^2 < \infty,$$

which proves (1).

The dual space of $(\mathscr{D}, \tau)$ is denoted by $\mathscr{D}'$ and an element of $\mathscr{D}'$ is called a (*Schwartz*) *distribution*. We can discuss $\mathscr{D}_p$, $\mathscr{D}'_p$ and $\|\cdot\|'_p$ in the same way as we discussed $\mathscr{S}_p$, $\mathscr{S}'_p$ and $\|\cdot\|'_p$ in the last section. By the identification of a function $f \in L_2$ with a distribution $\alpha_f = (f, \cdot)$ we obtain

$$\mathscr{D} \subset \mathscr{D}_p = \mathscr{D}'_{-p} \subset \mathscr{D}_0 = \mathscr{D}'_0 = L_2 \subset \mathscr{D}'_p \subset \mathscr{D}'$$

for $p > 0$, and $C^\infty[0, \pi] \supset \bigcap_p \mathscr{D}_p$, $\mathscr{D}' = \bigcup_p \mathscr{D}'_p$.

The d-dimensional case where $I = \prod_{i=1}^d [a_i, b_i]$ is dealt with similarly by using the Laplace operator $-\Delta$ for D and

$$s_{n_1 n_2 \cdots n_d}(t_1, t_2, \ldots, t_d) := \prod_{i=1}^d s_{n_i}(t_i)$$

for $s_n(t)$.

1.5. The spaces $\mathscr{D}$ and $\mathscr{D}'$ on $\mathbf{R}^d$. Let $\mathscr{D} = \mathscr{D}(\mathbf{R}^d)$ be the vector space of all C^∞-functions on $\mathbf{R}^d$ with compact support. Then

$$\mathbf{R}^d = \bigcup_n I_n^0, \qquad I^0 = \text{the interior of } I$$

where $\{I_n, \ n = 1, 2, \ldots\}$ is a rearrangement of $\prod_{i=1}^d [n_i - 1, n_i + 1]$, $n_i = 0, \pm 1, \pm 2, \ldots$. Let $\{\alpha_n\} \subset \mathscr{D}$ be a partition of unity corresponding to the covering $\{I_n^0\}$ of $\mathbf{R}^d$.

For every $\varphi \in \mathscr{D}$, $\alpha_n \varphi$ is viewed as a function belonging to the space $\mathscr{D}(I_n)$ introduced in the last section. Hence we can define a family of H-seminorms on $\mathscr{D}$:

$$\|\varphi\|_{n,p} := \|\alpha_n \varphi\|_p, \qquad n \in \mathbf{N}, \quad p \in \mathbf{R},$$

where $\{\|\cdot\|_p\}$ is the family of H-seminorms in $\mathscr{D}(I_n)$ introduced in the last section. Observing that $\|\cdot\|_p \prec_{HS} \|\cdot\|_{p+1}$, we can easily check that

$$\|\cdot\|_{n,p} \prec_{HS} \|\cdot\|_{n,p+1} \tag{1}$$

for every n.

For two sequences (a_n) and (p_n) such that $a_n > 0$ and $p_n = 0, 1, 2, \ldots$, we set

$$\|\varphi\|_{(a_n),(p_n)}^2 = \sum_n a_n^2 \|\varphi\|_{n,p_n}^2.$$

Since the support of $\varphi \in \mathscr{D}$ intersects only a finite number of the intervals $I_1, I_2, \ldots$, this norm is well defined and turns out to be an H-norm. The multi-Hilbertian topology $\tau = \tau(\mathscr{D})$ determined by the family $\{\|\cdot\|_{(a_n)(p_n)}\}$ coincides with the Schwartz topology in $\mathscr{D}$ [27]. From (1) we obtain

$$\|\cdot\|_{(a_n),(p_n)} \prec_{HS} \|\cdot\|_{(b_n),(p_n+1)} \qquad (b_n = na_n (\|\cdot\|_{n,p_n} : \|\cdot\|_{n,p_n+1})_{HS}),$$

so $I(\tau) = \tau$, namely τ is nuclear. It is obvious that $\|\cdot\|_{n,p} \prec \tau$, $n \in \mathbf{N}$, $p \in \mathbf{R}$. We can easily check that τ is not countably Hilbertian.

The dual space of $(\mathscr{D}, \tau)$ is called the space of *distributions*, written $\mathscr{D}' = \mathscr{D}'(\mathbf{R}^d)$. In the same way as in $\mathscr{S}_p$, $\mathscr{S}'_p$, etc., we can discuss $\mathscr{D}_{n,p}$, $\mathscr{D}'_{n,p}$, $\mathscr{D}_{(a_n),(p_n)}$, $\mathscr{D}'_{(a_n),(p_n)}$, etc. A locally integrable function f can be identified with $\alpha_f \in \mathscr{D}'$:

$$\alpha_f(\varphi) := \int_{\mathbf{R}^d} f\varphi \, dx.$$

In particular, by Sobolev's lemma or directly by using the Fourier series expansion we obtain the following fact.

If $\alpha \in \mathscr{D}'_{n,-p}$ *for every* $n = 1, 2, \ldots$ *and every* $p = 0, 1, 2, \ldots$, *then* α *is identified with a C^∞-function.*

Remark 1.5.1. From a tempered distribution $f \in \mathscr{S}'$ we obtain a distribution $f|_{\mathscr{D}} \in \mathscr{D}'$ (restriction to $\mathscr{D}$). Since $f \to f|_{\mathscr{D}}$ is injective, we can identify f with $f|_{\mathscr{D}}$, so $\mathscr{S}'$ is embedded into $\mathscr{D}'$. The topology $\tau(\mathscr{S})$, when restricted to $\mathscr{D}$, is weaker than the topology $\tau(\mathscr{D})$. $\mathscr{S}'$ is regarded as a vector subspace of $\mathscr{D}'$ consisting of all elements in $\mathscr{D}'$ continuous with respect to $\tau(\mathscr{S})|_{\mathscr{D}}$. Let $\|\cdot\|_p$ be the norm on $\mathscr{D}$ obtained by restricting the norm $\|\cdot\|_p$ on $\mathscr{S}$ to $\mathscr{D}$. Observing that

$$\|\varphi\|_p = \|D^p\varphi\| \left(D = x^2 - \frac{d^2}{dx^2} \right), \qquad p \in \mathbf{N},$$

we can find (a_n), (p_n) such that

$$\|\cdot\|_p \prec \|\cdot\|_{(a_n),(p_n)},$$

which proves that $\tau(\mathscr{S})|_{\mathscr{D}}$ is weaker than $\tau(\mathscr{D})$.

1.6. The spaces $\mathscr{D}$ and $\mathscr{D}'$ on a manifold. Let M be a σ-compact d-dimensional C^∞-manifold and

$$A = \{(x_n = (x_n^1, x_n^2, \ldots, x_n^d), U_n) : n = 1, 2, 3, \ldots\}$$

be an atlas defining the C^∞-structure of M, where every U_n is relatively compact, every compact subset of M intersects only a finite number of U_n, $n = 1, 2, \ldots$ and the closures of the images $x_n(U_n)$, $n = 1, 2, \ldots$, are compact intervals $I_n \subset \mathbf{R}^d$, $n = 1, 2, \ldots$. Let $\mathscr{D} = \mathscr{D}(M)$ be the vector space of all C^∞-functions on M with compact support. Let $\{\alpha_n\}$ be a partition of unity corresponding to the covering $\{U_n\}$ of M.

For $\varphi \in \mathscr{D}$ $\alpha_n\varphi$ has support in U_n. Hence we can regard it as a function of the coordinate $x_n \in I_n$, so that $\alpha_n\varphi \in \mathscr{D}(I_n)$. In the same way as in the last section we define $\|\cdot\|_{n,p}$, $\|\cdot\|_{(a_n),(p_n)}$, $\tau = \tau(\mathscr{D})$ and $\mathscr{D}'$. Then τ is multi-Hilbertian and nuclear and coincides with the Schwartz topology on $\mathscr{D}$ [27].

Let M_n be the closure of $\bigcup_{i=1}^{n} U_n$ and $\mathscr{D}(M_n)$ denote the set of all C^∞-functions on M with support $\subset M_n$. $\mathscr{D}(M_n)$ is a topological vector subspace of $\mathscr{D}(M)$, where the topology τ_n in $\mathscr{D}(M_n)$ is the relative topology induced from that in $\mathscr{D}(M)$. Since U_i is relatively compact for every i, M_n is compact. Hence M is covered by U_j, $j = 1, 2, \ldots, m$ for some $m < \infty$, so the topology in $\mathscr{D}(M_n)$ is determined by a countable family of H-norms:

$$\|\cdot\|_p = \bigvee_{j=1}^{m} \|\cdot\|_{j,p}, \qquad p = 0, 1, 2, \ldots.$$

Hence $\mathscr{D}(M_n) = (\mathscr{D}(M_n), \tau_n)$ is countably Hilbertian and nuclear. The dual space of $\mathscr{D}(M_n)$ is denoted by $\mathscr{D}'(M_n)$.

If M itself is compact, the atlas A consists of a finite number of charts. In this case τ is determined by a countable family

$$\|\cdot\|_p = \bigvee_{n} \|\cdot\|_{n,p}, \qquad p = 0, 1, 2, \ldots.$$

Hence τ is countably Hilbertian and nuclear.

We can prove that τ does not depend on the choice of the atlas A and the partition of unity $\{\alpha_n\}$. Hence τ has an intrinsic meaning.

If M is a Riemannian manifold, we have a natural measure μ induced from the metric tensor. μ is positive on opens and finite on compacts. Hence μ is σ-finite. Through this measure μ we can identify a locally integrable function f with the distribution

$$\alpha_f(\varphi) = \int_M f\varphi \, d\mu.$$

If $\alpha \in \mathscr{D}'_{n,-p}$ for every $n = 1, 2, \ldots$ and every $p = 0, 1, 2, \ldots$, then α is identified with a C^∞-function.

Suppose that M is not Riemannian. Fix a strictly positive C^∞-function ρ, for example $\rho \equiv 1$.

$$\mu(\varphi) = \sum_n \int_{I_n} \varphi \alpha_n \rho \, dx_n^1 \, dx_n^2 \cdots dx_n^d, \qquad \varphi \in \mathscr{D}(M),$$

defines a measure μ. Through this measure μ we can identify a function f with a distribution as above. This identification depends on the choice of A, $\{\alpha_n\}$ and ρ. But the property that a distribution α can be identified with a C^r-function ($r = 0, 1, 2, \ldots, \infty$) is independent of the choice. Hence the condition that a distribution α is a C^∞-function has an intrinsic meaning.

If $\alpha \in \mathscr{D}'_{n,-p}$ for every $n = 1, 2, \ldots$ and every $p = 0, 1, 2, \ldots$, then α is a C^∞-function.

Let $\mathscr{D}(p, q)$ be the vector space of all C^∞-tensor fields of type (p, q) on M with compact support. We can make the same discussion as above to introduce a multi-Hilbertian topology τ on $\mathscr{D}(p, q)$, which is also nuclear. The dual space of $(\mathscr{D}(p, q), \tau)$ is the space of currents of type (q, p), so it is denoted by $\mathscr{D}(q, p)'$. The facts observed above can be extended to this case.

CHAPTER 2

Infinite Dimensional Random Variables and Stochastic Processes

2.1. Standard measurable spaces. An abstract set S endowed with a σ-algebra $\mathscr{B}(S)$ of subsets of S (or simply called a σ-algebra on S) is called a *measurable space* $(S, \mathscr{B}(S))$. $\mathscr{B}(S)$ is often called the *measurable structure* of $(S, \mathscr{B}(S))$. For every subset $T \subset S = (S, \mathscr{B}(S))$,

$$\mathscr{B}(S) \cap T = \{B \cap T : B \in \mathscr{B}(S)\}$$

is a σ-algebra on T. T is regarded as a measurable space $(T, \mathscr{B}(S) \cap T)$, unless otherwise stated. $(T, \mathscr{B}(S) \cap T)$ is called a *measurable subspace* of $(S, \mathscr{B}(S))$. Let

$$S_i = (S_i, \mathscr{B}(S_i)), \qquad i \in I$$

be a family of measurable spaces and S their Cartesian product $\prod_{i \in I} S_i$. Let p_i be the natural projection of S to S_i carrying $s \in S$ to the ith component. The σ-algebra on S generated by the sets

$$p_i^{-1}(B_i), \qquad B_i \in \mathscr{B}(S_i), \quad i \in I$$

is called the *product σ-algebra* of $\mathscr{B}(S_i)$ $i \in I$, written $\prod_{i \in I} \mathscr{B}(S_i)$. This σ-algebra is called the *Kolmogorov σ-algebra* on $S = \prod_i S_i$ written $\mathscr{B}_K(S)$. The measurable space

$$(S, \mathscr{B}_K(S)) \equiv \left(\prod_{i \in I} S_i, \prod_{i \in I} \mathscr{B}(S_i) \right)$$

is called the *measurable product* of $(S_i, \mathscr{B}(S_i))$, $i \in I$.

A subset A of $S = (S, \mathscr{B}(S))$ is called *measurable* or *measurable* $\mathscr{B}(S)$, if $A \in \mathscr{B}(S)$. A map $f : S = (S, \mathscr{B}(S)) \to T = (T, \mathscr{B}(T))$ is called *measurable* or *measurable* $\mathscr{B}(S)/\mathscr{B}(T)$, written $f \in \mathscr{B}(S)/\mathscr{B}(T)$, if

$$f^{-1}(\mathscr{B}(T)) \subset \mathscr{B}(S)$$

where the left-hand side is defined to be $\{f^{-1}(B) : B \in \mathscr{B}(T)\}$.

Let S be a topological space. The σ-algebra generated by the open subsets of S is called the *topological σ-algebra* or the *Borel system* of S, denoted by $\mathscr{B}(S)$. Every topological space S is regarded as a measurable space endowed with the topological σ-algebra, unless otherwise stated. A subset of S is called a *Borel set* if it belongs to $\mathscr{B}(S)$, namely, if it is a measurable subset of $(S, \mathscr{B}(S))$. If T is a topological subspace of S, namely, if T is a subset of S endowed with the relative topology, then the space $T = (T, \mathscr{B}(T))$ is also a measurable subspace of S, because $\mathscr{B}(T) = \mathscr{B}(S) \cap T$. However, the topological product of a family of topological spaces S_i, $i \in I$ does not always coincide with the measurable product of these spaces. In general the topological σ-algebra on $S = \prod_i S_i$ is larger than the Kolmogorov σ-algebra $\mathscr{B}_K(S)$. If I is countable and if every S_i has a countable open base, then these two σ-algebras coincide with each other, so the topological product S, viewed as a measurable space, coincides with the measurable product $(S, \mathscr{B}_K(S))$.

$\mathbf{R}$, $\mathbf{R}^n$ ($n = 1, 2, \ldots, \infty$) and $\mathbf{R}^I$ and their subsets are topological spaces and so are regarded as measurable spaces. In general the topological σ-algebra $\mathscr{B}(\mathbf{R}^I)$ is larger than the product σ-algebra $\mathscr{B}_K(\mathbf{R}^I)$, and

$$\mathscr{B}_K(\mathbf{R}^I) = \mathscr{B}(\mathbf{R}^I)$$

if and only if I is countable. Let S be a subset of $\mathbf{R}^I$. Then the *Kolmogorov σ-algebra* on S, written $\mathscr{B}_K(S)$, is defined to be $\mathscr{B}_K(\mathbf{R}^I) \cap S$. $(S, \mathscr{B}_K(S))$ is a measurable subspace of $(\mathbf{R}^I, \mathscr{B}_K(\mathbf{R}^I))$.

Let S and T be measurable spaces. A map $f: S \to T$ is called *bimeasurable* if f is bijective and if both f and f^{-1} are measurable. T is said to be *Borel isomorphic* (abbr. B-isomorphic) to S, written $T \sim S$, if there exists a bimeasurable map $f: S \to T$. This relation is an equivalence relation.

DEFINITION 2.1.1. A measurable space is called a *standard measurable space*, if it is B-isomorphic to a Borel subset of $\mathbf{R}$ (viewed as a measurable space).

The following theorem is well known [11], though not elementary.

THEOREM 2.1.1. (i) *Every standard measurable space is B-isomorphic to one of the following standard spaces*:

$$I = [0, 1], \quad \mathbf{N}_n = \{1, 2, 3, \ldots, n\} \, (n = 1, 2, \ldots), \quad \mathbf{N}.$$

(ii) *Let S and T be standard measurable spaces. If $f: S \to T$ is a measurable injection, then $f(S)$ is Borel isomorphic to S (under f) and $f(S) \in \mathscr{B}(T)$.*

Let us mention some immediate consequences of this theorem. Since $\mathbf{N}$ is B-isomorphic to a compact subset of $\mathbf{R}$ (for example $\{0, 1, 1/2, 1/3, \ldots\}$), we obtain the following theorem:

THEOREM 2.1.2. *A measurable space is standard if and only if it is B-isomorphic to a compact subset of $\mathbf{R}$.*

Since B-isomorphism is preserved by measurable products and since the countable product of the measurable spaces mentioned in Theorem 2.1.1 (i) are B-isomorphic to one of these spaces, we obtain the following theorem:

THEOREM 2.1.3. *The measurable product of countably many standard measurable spaces is standard.*

Let $\{S_n, n = 1, 2, \ldots\}$ be a sequence of measurable subspaces of a measurable space S such that S is the (set-theoretic) union of S_n, $n = 1, 2, \ldots$. If $\{S_n\}$ is disjoint, then S is called the *measurable disjoint union* of $\{S_n\}$. If $\{S_n\}$ is increasing, then S is called the *measurable increasing limit* of $\{S_n\}$. In this case S is the *inductive limit* of $\{S_n\}$ with respect to inclusion maps $i_{m,n}: S_n \to S_m$ ($n < m$).

THEOREM 2.1.4. *The standard measurable spaces are closed under measurable disjoint unions and measurable increasing limits.*

Proof. Let S be the measurable disjoint union of standard measurable subspaces $S_1, S_2, \ldots$. Then each S_n is B-isomorphic to $B_n \in \mathscr{B}(n, n+1)$. Since

B_n, $n = 1, 2, \ldots$ are disjoint, $B := \bigcup_n B_n \in \mathscr{B}(\mathbf{R})$ is B-isomorphic to S. Hence S is standard.

Let S be the measurable increasing limit of standard measurable subspaces $S_1, S_2, \ldots$. Since

$$\mathscr{B}(S_n) = \mathscr{B}(S) \cap S_n = \mathscr{B}(S_{n+1}) \cap S_n,$$

the inclusion map $i_n : S_n \to S_{n+1}$ is measurable. Hence $S_n = i_n(S_n) \in \mathscr{B}(S_{n+1})$ by Theorem 2.1.1 (ii). Therefore $T_n := S_{n+1} - S_n \in \mathscr{B}(S_{n+1})$ ($-$ denotes the proper difference) and so T_n is standard. But S is the measurable disjoint union of S_1, $T_1, T_2, \ldots$, so S is standard. $\quad\square$

Let $\{S_n\}$ be a sequence of measurable spaces, and $p_{n,n+1} : S_{n+1} \to S_n$ be measurable for every n. Let S be a measurable space and $p_n : S \to S_n$ be measurable for every n. Suppose that

$$p_n = p_{n,n+1} \circ p_{n+1} \tag{1}$$

and

$$p_{n,n+1}(s_{n+1}) = s_n, n = 1, 2, \ldots \Rightarrow \exists_1 s : p_n(s) = s_n, n = 1, 2, \ldots \tag{2}$$

and that $\mathscr{B}(S)$ is generated by $\bigcup_n p_n^{-1}(\mathscr{B}(S_n))$. Then $S = (S, \mathscr{B}(S))$ is called the *projective limit* of $\{S_n\}$ with respect to $\{p_{n,n+1}, p_n\}_n$. In fact S is B-isomorphic to the following subspace T of the measurable product $U = \prod_n S_n$:

$$T = \{t = (t_1, t_2, \ldots) \in U : p_{n,n+1}(t_{n+1}) = t_n, n = 1, 2, \ldots\}$$

by the bimeasurable map

$$f : S \to T, \qquad s \mapsto (p_1(s), p_2(s), \ldots).$$

To prove the bimeasurability of f, use (1), (2) and the definition of $\mathscr{B}(S)$.

THEOREM 2.1.5. *The standard measurable spaces are closed under projective limits.*

Proof. Using the notation above, we will prove that S is standard if every S_n is standard. U is standard by Theorem 2.1.3. Since $T \in \mathscr{B}(U)$ by measurability of $p_{n,n+1}$, T is standard by Definition 2.1.1. Since f is bimeasurable, S is also standard. $\quad\square$

DEFINITION 2.1.2. A topological space S is called a *Polish space* if there exists a metric d defining the topology in S such that the metric space (S, d) is complete and separable.

THEOREM 2.1.6. *Every Polish space, viewed as a measurable space, is standard.*

Proof. It suffices to prove that a metric space (S, d) is a standard measurable space if it is complete and separable. Take a dense sequence $\{a_n\}$ in S and define

$$f : S \to \mathbf{R}^\infty, \qquad x \mapsto (d(x, a_n), n = 1, 2, \ldots).$$

Then it is easy to check that S is homeomorphic to $f(S)$. But the completeness of (S, d) implies that

$$f(S) = \bigcap_m \bigcup_r \bigcap_n \{\eta = (\eta_1, \eta_2, \ldots) : |\eta_n - d(a_r, a_n)| < m^{-1}\}$$

$$\in \mathscr{B}(\mathbf{R}^\infty).$$

As $\mathbf{R}^\infty$ is a standard measurable space by Theorem 2.1.3, so is $f(S)$. Since homeomorphism implies B-isomorphism, S is also standard. $\square$

Being a Polish space, every separable Banach space (and so every separable Hilbert space) is a standard measurable space, when it is endowed with the topological σ-algebra with respect to the norm topology.

Now we will prove that the spaces $\mathscr{S}'$ and $\mathscr{D}'$ observed in the previous section endowed with the Kolmogorov σ-algebras (§ 1.2) are standard measurable spaces.

The following spaces are dual countably Hilbertian spaces:

$\mathscr{S}'$ on $\mathbf{R}^d$ (§ 1.3),

$\mathscr{D}'$ on $[a, b]$ (§ 1.4),

$\mathscr{D}'$ and $\mathscr{D}'(p, q)$ on a compact manifold M (§ 1.6).

Hence these spaces, endowed with the Kolmogorov σ-algebras, are standard by virtue of the following theorem.

THEOREM 2.1.7. *Let $S = (S, \tau)$ be a countably Hilbertian space. Then $(S', \mathscr{B}_K(S'))$ is a standard measurable space.*

Proof. We can assume that τ is defined by an increasing sequence of H-seminorms $\{p_n\}$. Let S'_n be a vector subspace of S' consisting of all p_n-continuous linear functionals on S. S'_n is a separable Hilbert space with the dual norm p'_n. Let $\mathscr{B}(S'_n)$ be the topological σ-algebra on S' with respect to the p'_n-topology. Then $(S'_n, \mathscr{B}(S'_n))$ is standard, as we have mentioned above as an example of Theorem 2.1.6. But it is easy to check that $\mathscr{B}(S'_n) = \mathscr{B}_K(S'_n)$. Hence $(S'_n, \mathscr{B}_K(S'_n))$ is standard. But we have

$$S'_1 \subset S'_2 \subset \cdots, \qquad S' = \bigcup_n S'_n,$$

$$\mathscr{B}_K(S'_n) = \mathscr{B}_K(S') \cap S'_n,$$

so $(S', \mathscr{B}_K(S'))$ is the measurable increasing limit of $(S'_n, \mathscr{B}_K(S'_n))$, $n = 1, 2, \ldots$. Hence $(S', \mathscr{B}_K(S'))$ is standard by Theorem 2.1.4. $\square$

Each of the following spaces is the measurable projective limit of a sequence of dual countably Hilbertian spaces and so is standard by Theorems 2.1.7 and 2.1.4:

$$\mathscr{D}' \text{ on } \mathbf{R}^d \ (\S\ 1.5),$$

$\mathscr{D}'$ and $\mathscr{D}'(q, p)$ on a noncompact manifold M (§ 1.6).

In fact $(\mathscr{D}'(M), \mathscr{B}_K(\mathscr{D}'(M)))$ is the projective limit of $(\mathscr{D}'(M_n), \mathscr{B}_K(\mathscr{D}'(M_n)))$, $n = 1, 2, \ldots$ with respect to the measurable maps.

$$p_{n,n+1}(\alpha) = \alpha \big|_{\mathscr{D}(M_n)} \in \mathscr{D}'(M_n), \qquad \alpha \in \mathscr{D}'(M_{n+1}).$$

As $(\mathscr{D}'(M_n), \mathscr{B}_K(\mathscr{D}'(M_n)))$, $n = 1, 2, \ldots$, are standard, so is $(\mathscr{D}'(M), \mathscr{B}_K(\mathscr{D}'(M)))$, and similarly for other spaces.

2.2. Fundamental concepts in probability theory. Let f be a map of a set E into F. If $\mathscr{E}$ is a σ-algebra on E, then

$$f\mathscr{E} := \{B \subset F : f^{-1}(B) \in \mathscr{E}\}$$

is a σ-algebra on F, called the *image σ-algebra* of $\mathscr{E}$ by f. It is obvious that $f\mathscr{E}$ is different from the image $f(\mathscr{E}) := \{f(A) : A \in \mathscr{E}\}$; $f(\mathscr{E})$ is not always a σ-algebra on F.

Let E be a set. A $[0, 1]$-valued set function μ on E defined on a class of subsets of E, written $\mathscr{D}(\mu)$, is called a *probability measure* on E, if (i) $\mathscr{D}(\mu)$ is a σ-algebra on E and (ii) μ is σ-additive and $\mu(E) = 1$. A set E endowed with a probability measure μ is called a *probability space* (E, μ). A set $A \subset E$ is called μ-measurable if $A \in \mathscr{D}(\mu)$. *Completeness* and the *Lebesgue extension* of a probability measure are defined as usual.

Let μ be a probability measure on a set E and f a map of E into another set F. Then

$$\nu(B) = \mu(f^{-1}(B)), \qquad \mathscr{D}(\nu) = f\mathscr{D}(\mu)$$

defines a probability measure ν on F, called the *image measure* of μ by f. It is denoted by $f\mu$ or μf^{-1}. Completeness is inherited by image measures.

Let $(E, \mathscr{E})$ be a measurable space. A probability measure μ on E is called a *Borel probability measure* on $(E, \mathscr{E})$ if $\mathscr{D}(\mu) = \mathscr{E}$. The Lebesgue extension of a Borel probability measure on $(E, \mathscr{E})$ is called a *Lebesgue probability measure* on $(E, \mathscr{E})$. In other words a probability measure on E is called a Lebesgue probability measure on $(E, \mathscr{E})$, if (i) μ is complete, (ii) $\mathscr{D}(\mu) \supset \mathscr{E}$ and (iii) for every $A \in \mathscr{D}(\mu)$ we can find $B_1, B_2 \in \mathscr{E}$ such that

$$B_1 \subset A \subset B_2 \quad \text{and} \quad \mu(B_2 \backslash B_1) = 0.$$

Let (E, μ) be a probability space and $(F, \mathscr{F})$ a measurable space. A map $f : E \to F$ is called a *μ-measurable function* on (E, μ) if f is measurable $\mathscr{D}(\mu)/\mathscr{F}$. In this case $\mathscr{D}(f\mu)$ includes $\mathscr{F}$. (E, μ) or μ is called *perfect* if (i) μ is complete and (ii) for every μ-measurable $(\mathbf{R}, \mathscr{B}(\mathbf{R}))$-valued function f on (E, μ), the image measure $f\mu$ is a Lebesgue (*not in the classical sense but in the sense above*) probability measure on $(\mathbf{R}, \mathscr{B}(\mathbf{R}))$.

THEOREM 2.2.1. *Every Lebesgue probability measure on a compact Hausdorff space is perfect.*

Proof. Let K be a compact Hausdorff space, $\mathscr{B}(K)$ the topological σ-algebra on K and μ a Lebesgue probability measure on $(K, \mathscr{B}(K))$. For every μ-measurable set and every $\varepsilon > 0$ we can find a compact set $C \subset A$ such that (i) $\mu(A \backslash C) < \varepsilon$ and (ii) the restriction of f to C is continuous, so $f(C)$ is compact (a generalization of Lusin's theorem). Using this fact we can easily prove this theorem. $\square$

THEOREM 2.2.2. *Every Lebesgue measure on a standard measurable space is perfect.*

Proof. Use the last theorem, observing that every standard measurable space is B-isomorphic to a compact subset of $\mathbf{R}$ (Theorem 2.1.2). $\square$

THEOREM 2.2.3. *Let (E, μ) be a perfect probability space. If f is a μ-measurable* *function on (E, μ) with values in a standard measurable space $(F, \mathscr{F})$, then the image measure $f\mu$ is a Lebesgue probability measure on $(F, \mathscr{F})$.*

Proof. It suffices to recall the definition of a standard measurable space. $\square$

In probability theory we fix a probability space (Ω, P) and call a P-measurable $(E, \mathscr{E})$-valued function an $(E, \mathscr{E})$-valued *random variable* on (Ω, P). Let X be a random variable. The image measure XP is called the *probability distribution* of X, denoted by P^X, in view of the fact that

$$B \in \mathscr{D}(P^X) \Leftrightarrow X^{-1}(B) \in \mathscr{D}(P) \Rightarrow (XP)(B) = P(X^{-1}(B)).$$

In this note we assume that (Ω, P) is perfect and observe only random variables with values in standard measurable spaces, unless stated otherwise. Then the probability distribution of a random variable is a perfect Lebesgue probability measure by the last theorem.

Let $\{X_\alpha, \alpha \in A\}$ be a family of random variables on (Ω, P), where each X_α takes values in a standard measurable space $(E_\alpha, \mathscr{E}_\alpha)$. For every $F \subset A$ we set

$$(E_F, \mathscr{E}_F) = \left(\prod_{\alpha \in F} E_\alpha, \prod_{\alpha \in F} \mathscr{E}_\alpha \right) \tag{1}$$

and denote by $\pi_{F,G}$ $(F \subset G \subset A)$ the natural projection of E_G to E_F. $\pi_{F,G}$ is measurable $\mathscr{E}_F/\mathscr{E}_G$. Let $\mathbf{F}$ be the class of all finite subsets of A. Then for every $F \in \mathbf{F}$, $(E_F, \mathscr{E}_F)$ is standard, and the joint variable

$$X_F(\omega) := (X_\alpha(\omega), \alpha \in f)$$

is a random variable with values in $(E_F, \mathscr{E}_F)$. Let μ_F denote the probability distribution of X_F, called the *joint distribution* of X_α, $\alpha \in F$. μ_F is a Lebesgue probability measure on $(E_F, \mathscr{E}_F)$ for every $F \in \mathbf{F}$. Since $X_F = \pi_{F,G} \circ X_G$, we have

$$(\mathrm{K}) \qquad\qquad \mu_F = \pi_{F,G}\mu_G, \quad F \subset G, \quad F, G \in \mathbf{F}.$$

This condition is called the *Kolmogorov consistency condition.*

THEOREM 2.2.4 (Kolmogorov's extension theorem). *Suppose that $(E_\alpha, \mathscr{E}_\alpha)$, $\alpha \in A$, are standard measurable spaces and define $(E_F, \mathscr{E}_F)$, $F \in \mathbf{F}$ by (1), where $\mathbf{F}$ is the class of all finite subsets of A. If μ_F is a Lebesgue measure on $(E_F, \mathscr{E}_F)$ for every $F \in \mathbf{F}$ and if the family $\{\mu_F, F \in \mathbf{F}\}$ satisfies the Kolmogorov consistency condition, then we can construct a probability space (Ω, p) and a family of random variables X_α, $\alpha \in A$ on (Ω, P) such that the joint distribution of $(X_\alpha, \alpha \in F)$ is μ_F for every $F \in \mathbf{F}$.*

Proof. Since $(E_\alpha, \mathscr{E}_\alpha)$ is standard, there exists a bimeasurable map f_α of E_α to a compact set $K_\alpha \subset \mathbf{R}$ by Theorem 2.1.2. Then the image measure ν_F of μ_F by the map

$$f_F(x_\alpha, \alpha \in F) = (f_\alpha(x_\alpha), \alpha \in F)$$

is a Lebesgue probability measure on $(K_F, \mathscr{B}(K_F))$ for $F \in \mathbf{F}$, where

$$K_F := \prod_{\alpha \in F} K_\alpha.$$

K_F and $\Omega := \prod_{\alpha \in A} K_\alpha$ are compact Hausdorff spaces by Tikhonov's theorem. Let $C(\Omega)$ be the Banach space of all continuous functions with the maximum norm $\|\cdot\|$ and $C_1(\Omega)$ the subspace of $C(\Omega)$ that consists of all continuous tame functions on Ω, namely of all functions f's represented in the form

$$f = f_F \circ \pi_{F,A}, \qquad f_F \in C(K_F), \quad F \in \mathbf{F}.$$

$C_1(\Omega)$ is dense in $C(\Omega)$.

Define a linear functional L on $C_1(\Omega)$ by

$$L(f) = \int_{K_F} f_F \, d\nu_F, \qquad f = f_F \circ \pi_{F,A}.$$

$L(f)$ is well defined independently of the representation $f = f_F \circ \pi_{F,A}$ by the consistency condition. It is obvious that $L(f) \geqq 0$ for $f \geqq 0$ and $L(1) = 1$, so $|L(f)| \leqq \|f\|$. Since $C_1(\Omega)$ is dense in $C(\Omega)$, we can extend L to a bounded linear functional on $C(\Omega)$ having the same properties. By the Riesz representation theorem we obtain a Lebesgue probability measure P on $(\Omega, \mathscr{B}(\Omega))$ such that $L(f)$ equals the P-integral of f for every $f \in C(\Omega)$. P is perfect by Theorem 2.2.1.

Denote the natural projection of Ω to K_α by Y_α. Then $Y_\alpha : \Omega \to K_\alpha$ is measurable $\mathscr{B}(\Omega)/\mathscr{B}(K_\alpha)$. Hence Y_α is a $(K_\alpha, \mathscr{B}(K_\alpha))$-valued random variable. Let Y_F be the joint variable of Y_α, $\alpha \in F$ for $F \in \mathbf{F}$. Then Y_F is a $(K_F, \mathscr{B}(K_F))$-valued random variable on (Ω, P). By the construction of P we can easily see that the probability distribution of Y_F is ν_F.

Since $f_\alpha : E_\alpha \to K_\alpha$ is bimeasurable, its inverse map $g_\alpha : K_\alpha \to E_\alpha$ is measurable $\mathscr{B}(K_\alpha)/\mathscr{E}_\alpha$. Hence $X_\alpha := g_\alpha \circ Y_\alpha$ is an $(E_\alpha, \mathscr{E}_\alpha)$-valued random variable on (Ω, P). Then $g_F(y_\alpha, \alpha \in F) := (g_\alpha(y_\alpha), \alpha \in F)$ is an inverse map of f_F measurable $\mathscr{E}_F/\mathscr{B}(K_F)$ for $F \in \mathbf{F}$, the joint variable $X_F := (X_\alpha, \alpha \in F)$ is equal to $g_F \circ Y_F$. Hence the probability distribution of X_F is equal to the probability measure

$$g_F \nu_F = g_F(f_F \mu_F) = \mu_F. \qquad\qquad \Box$$

Remark 2.2.1. In the proof above we used $\mathscr{B}(K_F) = \prod_{\alpha \in F} \mathscr{B}(K_\alpha)$, which is verified by observing that $K_\alpha \subset \mathbf{R}$ and F is a finite set.

Let $\mathscr{F}$ be a σ-subalgebra of $\mathscr{D}(P)$ and $X = X(\omega)$ be an $(E, \mathscr{E})$-valued random variable.

THEOREM 2.2.5. *There exists a function* $\mu(B, \omega)$, $B \in \mathscr{E}$, $\omega \in \bar{\Omega}$ *satisfying the following conditions*:

(i) *For every* $\omega \in \Omega$, $\mu(B, \omega)$, *viewed as a function of* $B \in \mathscr{E}$, *is a probability measure on* $(E, \mathscr{E})$.

(ii) *For every* $B \in \mathscr{E}$, $\mu(B, \omega)$ *is* $\mathscr{F}$*-measurable in* ω.

(iii) *For every* $F \in \mathscr{F}$ *and every* $B \in \mathscr{E}$, $\int_F \mu(B, \omega) P(d\omega) = P\{X^{-1}(B) \cap F\}$. *Such a function* $\mu(B, \omega)$ *is uniquely determined up to* P*-measure* 0, *namely if* $\mu_1(B, \omega)$ *and* $\mu_2(B, \omega)$ *satisfy these conditions, then*

$$P\{\mu_1(B, \omega) = \mu_2(B, \omega), B \in \mathscr{E}\} = 1.$$

The function $\mu(B, \omega)$ is called the *conditional probability distribution* of X under $\mathscr{F}$, written $P^X(B \mid \mathscr{F})$.

Proof. This theorem obviously holds in the case $(E, \mathscr{E}) = (\mathbf{R}, \mathscr{B}(\mathbf{R}))$, and so in the case $E \in \mathscr{B}(\mathbf{R})$ and $\mathscr{E} = \mathscr{B}(\mathbf{R}) \cap E$. Therefore it holds for a standard measurable space $(E, \mathscr{E})$. $\square$

Let $L_0(\Omega) = L_0(\Omega, P)$ be the real random variables on (Ω, P). Let $X, Y \in L_0(\Omega)$. Y is said to be *equivalent* to X if $P(X = Y) = 1$, namely if $X = Y$ a.s. In the usual way we identify equivalent random variables in $L_0(\Omega)$. Then $L_0(\Omega)$ is a Fréchet space with quasi-norm

$$\|X\|_0 := E(|x| \wedge 1) = \int_\Omega |X(\omega)| \, P(d\omega).$$

This quasi-norm induces the topology of convergence in probability, because

$$\varepsilon P(|X| > \varepsilon) \leqq \|X\|_0 \leqq P(|X| > \varepsilon) + \varepsilon.$$

In $L_0(\Omega)$ we define $\|X\|_p$ for $p \in [1, \infty]$ as follows:

$$\|X\|_p := E(|X|^p)^{1/p}, \qquad p \in [I, \infty),$$

$$\|X\|_\infty := \operatorname{ess\,sup} |X| = \inf \{s : |X| \leqq s \text{ a.s.}\}.$$

Then

$$L_p(\Omega) = L_p(\Omega, P) = \{X \in L_0(\Omega) : \|X\|_p < \infty\}$$

is a Banach space with norm $\|\cdot\|_p$ for $p \in [1, \infty)$, and a Fréchet space for $p = 0$.

If Ω is a standard measurable space and P is a Lebesgue measure, then $L_p(\Omega)$ is separable for $p < \infty$, but $L_\infty(\Omega)$ is not always separable.

Since $L_p(\Omega)$ ($p \in \{0\} \cup [1, \infty]$) is a Banach space, we can define the topological σ-algebra $\mathscr{B}(L_p(\Omega))$. Except in the case $p = \infty$ the measurable space $(L_p(\Omega), \mathscr{B}(L_p(\Omega)))$ is standard. This is not always true for $p = \infty$.

Every random variable to be observed in this note takes values in the following standard measurable spaces and their variations.

  1. $(\mathbf{R}^n, \mathscr{B}(\mathbf{R}^n))$, $n = 1, 2, \ldots, \infty$.
  2. $(C, \mathscr{B}_K(C))$, $C = C(\mathbf{R}^n)$.

$\mathscr{B}_K(C)$ is the Kolmogorov σ-algebra on C, i.e. the σ-algebra generated by the sets

$$\{f \in C : f(t) < a\}, \qquad t \in \mathbf{R}^n, \quad a \in \mathbf{R}.$$

$\mathscr{B}_K(C)$ coincides with the topological σ-algebra $\mathscr{B}(C)$ with respect to the topology of uniform convergence on compacts. The space endowed with this topology is a Polish space. Hence $(C, \mathscr{B}_K(C) \equiv \mathscr{B}(C))$ is a standard measurable space by Theorem 2.1.6.

  3. $(D, \mathscr{B}_K(D))$, $D = D[\mathbf{R}]$.

D is the space of all right continuous functions on $\mathbf{R}$ with left limits. The space D with the Skorokhod topology is Polish [2], so the space D endowed with the topological σ-algebra $\mathscr{B}(D)$ is standard by Theorem 2.1.6. But $\mathscr{B}(D)$ coincides with $\mathscr{B}_K(D)$. Hence $(D, \mathscr{B}_K(D))$ is standard.

4. $(L_p, \mathcal{B}(L_p))(1 \leq p < \infty$, $L_p = L_p(\mathbf{R}^d)$, $\mathcal{B}(L_p) =$ the topological σ-algebra with respect to the norm topology), $(\mathcal{S}', \mathcal{B}_K(\mathcal{S}'))$, $(\mathcal{D}', \mathcal{B}_K(\mathcal{D}'))$.

These measurable spaces are standard as we have explained in § 2.1.

As we mentioned above, two real random variables X_1 and X_2 are said to be equivalent to each other if $P\{X_1 = X_2\} = 1$. We can extend this notion to random variables X_1 and X_2 taking values in standard measurable spaces $(S_1, \mathcal{B}(S_1))$ and $(S_2, \mathcal{B}(S_2))$ respectively. Equivalent random variables can be identified with each other.

Let us mention some examples of identification. Since every $f \in C \equiv C(\mathbf{R}^d)$ is locally integrable, we can identify f with $\alpha_f \in \mathcal{D}' \equiv \mathcal{D}'(\mathbf{R}^d)$:

$$\alpha_f(\varphi) := \int_{\mathbf{R}^d} f(x)\varphi(x)\, dx.$$

Thus C is regarded as a subset of $\mathcal{D}'$. It is easy to see that the inclusion map $i : C \to \mathcal{D}'$ is measurable $\mathcal{B}_K(C)/\mathcal{B}_K(\mathcal{D}')$, because $\mathcal{B}_K(C) = \mathcal{B}(C)$. Hence $C \in \mathcal{B}_K(\mathcal{D}')$ by Theorem 2.1.1(ii), and $(C, \mathcal{B}_K(\mathcal{D}') \cap C)$ coincides with $(C, \mathcal{B}_K(C))$. Hence it is obvious that every $(C, \mathcal{B}_K(C))$-valued random variable $X(\omega)$ is regarded as a $(\mathcal{D}', \mathcal{B}_K(\mathcal{D}'))$-valued random variable with $P(X \in C) = 1$. Conversely, if X is a $(\mathcal{D}, \mathcal{B}_K(\mathcal{D}'))$-valued random variable such that $P(X \in C) = 1$, then

$$X_1(\omega) = \begin{cases} X(\omega) & \text{if } X(\omega) \in C, \\ \text{any fixed element of } C & \text{otherwise} \end{cases}$$

is a $(C, \mathcal{B}_K(C))$-valued random variable. Since $P(X = X_1) = 1$, we can identify X with a $(C, \mathcal{B}_K(C))$-valued random variable X_1, and similarly for $(D, \mathcal{B}_K(D))$-, $(L_p, \mathcal{B}(L_p))$- and $(\mathcal{S}', \mathcal{B}_K(\mathcal{S}'))$-valued random variables.

Suppose that $P(X_i = Y_i) = 1$ for $i \in I$. If I is countable, then

$$P\{X_i(\omega) = Y_i(\omega), \, i \in I\} = 1.$$

Hence we can identify X_i with Y_i for $i \in I$ simultaneously. If I is not countable, then the simultaneous identification may not be possible in general.

2.3. Infinite dimensional random variables. In § 2.2 we defined an $(S, \mathcal{B}(S))$-valued random variable X to be an S-valued function on Ω measurable $\mathcal{D}(P)/\mathcal{B}(S)$. $S = (S, \mathcal{B}(S))$ is called the *sample space* of X. If S is a subset of $\mathbf{R}^d$ $(d < \infty)$ and $\mathcal{B}(S) = \mathcal{B}(\mathbf{R}^d) \cap S$, then X is a d-dimensional random variable. If S is a subset of the function space $\mathbf{R}^E$ and $\mathcal{B}(S) = \mathcal{B}_K(\mathbf{R}^E) \cap S$, then we may call X an *infinite dimensional random variable*. But such a general notion is not very useful, so we will restrict ourselves to the case where E is a multi-Hilbertian space and S is its dual space. Even under this restriction we can cover the cases

$$S = \mathcal{S}' \quad \text{and} \quad S = \mathcal{D}'.$$

The space $C = C(\mathbf{R}^d)$ is regarded as a measurable subset of $(\mathcal{D}', \mathcal{B}_K(\mathcal{D}'))$; the topological σ-algebra on C coincides with $\mathscr{B}_K(C)$ and $\mathscr{B}_K(C) = \mathscr{B}_K(\mathscr{D}') \cap C$. Hence every C-valued random variable X is regarded as a $\mathscr{D}'$-valued random variable satisfying $P(X \in C) = 1$, and similarly for $D(\mathbf{R})$-valued random variables. The restricted countable product of copies of $\mathbf{R}$,

$$\mathbf{R}_0^\infty = \{f \in \mathbf{R}^\infty : f_k = 0 \text{ except for finitely many } k\text{'s}\},$$

is countably Hilbertian with H-seminorms

$$p_n(f) = \left(\sum_{k=1}^n f_k^2 \right)^{1/2}, \qquad n = 1, 2, \ldots,$$

and its dual space is $\mathbf{R}^\infty$. Hence $(\mathbf{R}^\infty, \mathscr{B}_K(\mathbf{R}^\infty))$-valued random variables, namely *random sequences*, lie in our framework.

Hereafter $E_\tau = (E, \tau)$ always denotes a multi-Hilbertian space and an E_τ'-valued random variable means an $(E_\tau', \mathscr{B}_K(E_\tau'))$-valued random variable.

Let E be an abstract set. A family of random variables $Y = (Y_f, f \in E)$ is called a *random functional* on E. Y is regarded as a map

$$Y : E \to L_0(\Omega), \qquad f \mapsto Y_f,$$

and also as an $(\mathbf{R}^E, \mathscr{B}_K(\mathbf{R}^E))$-valued random variable,

$$Y(\omega) := (Y_f(\omega), f \in E_\tau),$$

called the *sample functional* of Y. This random variable $Y(\omega)$ is outside of our scheme in general.

Suppose that E is a vector space. If

$$h = af + bg \Rightarrow Y_h = aY_f + bY_g \text{ a.s.} \tag{1}$$

then $Y = \{Y_f, f \in E\}$ is called a *linear random functional* on E, where it should be noted that the exceptional P-null set for (1) may depend on a, b, f and g. (1) means that $Y : E \to L_0(\Omega)$ is linear.

Suppose that E is a topological space with topology τ. If

$$f \to f_0 \text{ (in } \tau) \Rightarrow Y_f \to Y_{f_0} \text{ (in } \|\cdot\|_0), \tag{2}$$

then $Y : E \to L_0(\Omega)$ is called τ-*continuous*.

Suppose that $E_\tau \equiv (E, \tau)$ is multi-Hilbertian. Even if $Y = \{Y_f, f \in E\}$ is a τ-continuous linear random functional on E, the sample functional $Y(\omega) = (Y_f(\omega), f \in E)$ is not always an E_τ'-valued random variable.

DEFINITION 2.3.1. An E_τ'-valued random variable X is called σ-*concentrated* if there exists a countably Hilbertian topology $\theta < \tau$ such that $P(X \in E_\theta') = 1$.

Let X be an E_τ'-valued random variable. Then $X(\omega)(f)$, viewed as a function of ω, is a real random variable, which is denoted by $X(f)$. Then $\{X(f), f \in E\}$ is a random functional. It is clearly linear. If X is σ-concentrated, then the random functional $X(f)$ is τ-continuous in f. To prove this, take a countably Hilbertian topology θ mentioned in the definition above. Since θ is weaker than τ, it suffices to prove that $X(f)$ is θ-continuous in f. Since θ is countably Hilbertian, E_θ is separable. Hence, for the proof of θ-continuity of $X(f)$ it suffices to observe that

$$f_n \to f \text{ in } \theta$$
$$\Rightarrow X(f_n) \to X(f) \quad \text{a.s. by } P(X \in E'_\theta) = 1$$
$$\Rightarrow \|X(f_n) - X(f)\|_0 \to 0 \quad \text{by the bounded convergence theorem.}$$

The following theorem gives a stronger result.

THEOREM 2.3.1. *If X is a σ-concentrated E'_τ-valued random variable, then $\{X(f), f \in E\}$ is linear and $I(\tau)$-continuous, where $I(\tau)$ is the Kolmogorov I-topology of τ introduced in § 1.2.*

Proof. We will use Sazonov's technique [26]. Take the topology θ mentioned in the definition above. Then θ is determined by an increasing sequence of H-seminorms $\{q_n\}$. Then

$$E'_n := E'_{q_n} \uparrow E'_\theta \quad \text{and so} \quad P(X \in E'_n) \uparrow 1. \tag{3}$$

E'_n is a separable Hilbert space with norm the dual norm of q_n, written q'_n. Let $\{e_k\}$ be an ONB in E_{q_n}. Then

$$\sum_k (X(\omega)(e_k))^2 = q'_n(X(\omega))^2 < \infty \quad \text{if } X(\omega) \in E'_n.$$

Let $\varepsilon > 0$. We can use (3) to find $n = n(\varepsilon)$ such that

$$P\left(\sum_k X(e_k)^2 < \infty\right) > 1 - \varepsilon,$$

so we can take $r = r(\varepsilon)$ such that

$$P\left(\sum_k X(e_k)^2 < r\right) > 1 - \varepsilon.$$

This implies that

$$\|X(f)\|_0^2 \leq E(X(f)^2 \wedge 1) \leq E\left(X(f)^2, \sum_k X(e_k)^2 < r\right) + \varepsilon. \tag{4}$$

Define $p(f) = p_\varepsilon(f)$ to be the square root of the $E(\cdot, \cdot)$ above. Then p is an H-seminorm. Since

$$\sum_k p(e_k)^2 = E\left(\sum_k X(e_k)^2, \sum_k X(e_k)^2 < r\right) < r,$$

p is a separable H-seminorm $<_{HS} q_n$. Hence

$$p < I(\theta) < I(\tau).$$

Since $\|X(f)\|_0^2 \leq p(f)^2 + \varepsilon$ by (4), we obtain

$$p(f) < \varepsilon \Rightarrow \|X(f)\|_0^2 < \varepsilon^2 + \varepsilon,$$

which proves that $X(f)$ is $I(\tau)$-continuous in f. $\square$

DEFINITION 2.3.2 [16]. Let τ be a multi-Hilbertian topology on a vector space E and $Y = \{Y(f)\}$ a linear random functional on E. An E'_τ-valued random variable X is called a τ-*regular version* of Y if (i) X is σ-concentrated and (ii) $X(f) = Y(f)$ a.s. for every $f \in E$.

If X is a τ-regular version of Y, the linear random functional $\{X(f), f \in E\}$ is $I(\tau)$-continuous by Theorem 2.3.1. Since $X(f) = Y(f)$ a.s. for every $f \in E$, $\{Y(f), f \in E\}$ is also $I(\tau)$-continuous. The converse is also true:

THEOREM 2.3.2. *If Y is an $I(\tau)$-continuous linear random functional on E, then there exists a τ-regular version of Y, which is unique in the sense that if X and X' are τ-regular versions of Y, then $X = X'$ a.s.*

Proof. Take a sequence $\varepsilon_n > 0$ such that

$$\sum_n \varepsilon_n > \infty.$$

Since $Y : E \to L_0(\Omega)$ is linear and $I(\tau)$-continuous, we can find $\delta_n > 0$ and $p_n < I(\tau)$ such that

$$p_n(f) < \delta_n \Rightarrow \|Y(f)\|_0 < \varepsilon_n. \tag{5}$$

By replacing p_n by $p_1 \vee p_2 \vee \cdots \vee p_n$ if necessary, we can assume that p_n increases with n. Relation (5) implies that

$$\mathrm{Re}\, E(e^{iY(f)}) > 1 - 2\varepsilon_n - 2p_n(f)^2 \delta_n^{-2}. \tag{6}$$

In fact, if $p_n(f) < \delta_n$, then

$$|E(e^{iY(f)}) - 1| \leq E(|Y(f)| \wedge 2) \leq 2\|Y(f)\|_0 < 2\varepsilon_n,$$

so the left-hand side of (6) is larger than $1 - 2\varepsilon_n$, and if $p_n(f) \geq \delta_n$, then the right-hand side of (6) is less than -1, whereas the left-hand side is no less than -1.

Denote by α the countably Hilbertian topology determined by $\{p_n\}$. Since $p_n < I(\tau)$, we can find $q_n < \tau$ such that $p_n <_{HS} q_n$. We can assume that $\{q_n\}$ is increasing. Let β be the countably Hilbertian topology determined by $\{q_n\}$. Then E has a countable β-dense subset

$$D = \{d_1, d_2, \ldots\}.$$

Since α is weaker than β, D is also α-dense in E.

Applying the Schmidt q_n-orthogonalization to D, we obtain a q_n-ONB $\{e_{kn}, k = 1, 2, \ldots\}$ on E such that d_j is expressible in the form

$$d_j = \sum_{k=1}^{j} \alpha_{jkn} e_{kn} + r_{jn}, \tag{7}$$

where

$$q_n(r_{jn}) = 0. \tag{8}$$

Since $p_n <_{HS} q_n$, this obviously implies that

$$p_n(r_{jn}) = 0. \tag{8'}$$

Fix n for the moment and suppress the index n in p_n and r_{jn} for simplicity of notation. Putting

$$f = \sum_{j=1}^{J} a_j r_j$$

in (6) and noting that $p(f) = 0$ by (8′), we obtain

$$\mathrm{Re}\, E\!\left(\exp\left(i \sum_{j=1}^{J} a_j Y(r_j)\right)\right) > 1 - 2\varepsilon.$$

Integrating both sides with respect to the measure $\prod_{j=1}^{J} N_v(da_j)$ where N_v is the centered Gaussian measure on $\mathbf{R}$ with variance $v > 0$, we obtain

$$E\!\left(\exp\left(-\frac{v}{2} \sum_{j=1}^{J} Y(r_j)^2\right)\right) > 1 - 2\varepsilon.$$

Letting $J \to \infty$ and then $v \to \infty$, we obtain

$$P\left\{\sum_{j=1}^{\infty} Y(r_j)^2 = 0\right\} \geq 1 - 2\varepsilon. \tag{9}$$

Setting

$$f = \sum_{k=1}^{K} a_k e_k$$

in (6), we obtain

$$\mathrm{Re}\, E\!\left(\exp\left(i \sum_{k=1}^{K} a_k Y(e_k)\right)\right) > 1 - 2\varepsilon - 2\delta^{-2} \sum_{j,k=1}^{K} a_k a_j p(e_k, e_j).$$

Integrating both sides with respect to the same measure as above, we obtain

$$E\!\left(\exp\left(-\frac{v}{2} \sum_{k=1}^{K} Y(e_k)^2\right)\right)$$

$$> 1 - 2\varepsilon - 2v\delta^{-2} \sum_{k=1}^{K} p(e_k)^2 \geq 1 - 2\varepsilon - 2v\delta^{-2}(p:q)^2_{\mathrm{HS}},$$

where this HS-ratio is finite by $p <_{\mathrm{HS}} q$. Letting $K \to \infty$ and $v \downarrow 0$, we obtain

$$P\left\{\sum_{k=1}^{\infty} Y(e_k)^2 < \infty\right\} \geq 1 - 2\varepsilon. \tag{10}$$

Now, restoring the index n, we can deduce from (9) and (10) that

$$P(\Omega_n) \geq 1 - 4\varepsilon_n, \qquad n = 1, 2, \ldots, \tag{11}$$

where

$$\Omega_n = \left\{Y(r_{jn}) = 0,\, j = 1, 2, \ldots,\, \sum_{k=1}^{\infty} Y(e_{kn})^2 < \infty\right\}. \tag{12}$$

Let $\{e'_{kn}\}_k$ be the dual ONB of $\{e_{kn}\}_k$ on E'_{qn} and define

$$X_n(\omega) := \begin{cases} \sum_k Y(e_{kn})(\omega)e'_{kn} & \text{on } \Omega_n, \\ 0 & \text{elsewhere.} \end{cases} \tag{13}$$

This random variable is well defined by virtue of (12). It is obvious that

$$X_n(\omega) \in E'_{q_n} \subset E'_\beta \quad \text{for every } \omega. \tag{14}$$

The expression (13) implies that

$$X_n(e_{kn}) = Y(e_{kn}), \qquad k = 1, 2, \ldots \quad \text{on } \Omega_n. \tag{15}$$

Since

$$|e'_{kn}(r_{jn})| \leq q'_n(e'_{kn})q_n(r_{jn}) = 0 \quad \text{(by (8))},$$

(12) and (13) imply that

$$X_n(r_{jn}) = Y(r_{jn}) = 0, \qquad j = 1, 2, \ldots \quad \text{on } \Omega_n. \tag{16}$$

But d_j is a linear combination of $\{e_{1n}, e_{2n}, \ldots e_{jn}, r_{jn}\}$ by (7), $X_n(f)$ is linear in f for every ω, and $Y : E \to L_0(\Omega)$ is linear. Hence (15) and (16) imply that

$$X_n(d_j) = Y(d_j), \qquad j = 1, 2, \ldots \quad \text{on } \Omega'_n := \Omega_n - N_n, \tag{17}$$

where $P(N_n) = 0$. Therefore

$$P(\Omega'_n) = P(\Omega_n) \geq 1 - 4\varepsilon_n$$

by (11). Since $\sum_n \varepsilon_n < \infty$, the Borel–Cantelli lemma implies that

$$P(\Omega') = 1, \quad \text{where } \Omega' := \bigcup_N \bigcap_{n \geq N} \Omega'_n. \tag{18}$$

By (17) and (18) we can find $N(\omega)$ such that

$$X_m(d_j) = X_n(d_j) = Y(d_j), \qquad j = 1, 2, \ldots \tag{19}$$

whenever $\omega \in \Omega'$ and $m, n \geq N(\omega)$. Since $\{d_j\}$ is β-dense in E and $X_k \in E'_\beta$ for every k, we can deduce from (19) that

$$X_m(f) = X_n(f), \qquad f \in E$$

whenever $\omega \in \Omega'$ and $m, n \geq N(\omega)$. Define

$$X(f) := \begin{cases} X_n(f), & \omega \in \Omega', n \geq N(\omega), \\ 0 & \text{elsewhere.} \end{cases} \tag{20}$$

It is obvious that $X \in E'_\beta \subset E'$ for every ω.

It remains only to prove that

$$X(f) = Y(f) \quad \text{a.s.}$$

Since $X_n(d_j) = Y(d_j)$ on Ω'_n and $P(\Omega'_n) \to 1$,

$$X_n(d_j) \to Y(d_j) \quad \text{i.p.}$$

Since $p(\Omega') = 1$, $X_n(d_j) \to X(d_j)$ a.s. by (20). Hence

$$X(d_j) = Y(d_j) \quad \text{a.s.} \tag{21}$$

Let f be any point in E. Since $\{d_j\}$ is β-dense in E, we can find a sequence $\{f_n\} \subset \{d_j\}$ that β-converges to f. Since $X \in E'_\beta$ for every ω,

$$X(f_n) \to X(f) \quad \text{for every } \omega.$$

Since α is weaker than β, $\{f_n\}$ α-converges to f, so $\|Y(f_n) - Y(f)\|_0 \to 0$ by (5). Since $X(f_n) = Y(f_n)$ a.s. by (21), $X(f) = Y(f)$ a.s. This proves the existence part of the theorem.

The uniqueness part can easily be proved by noting that there exists a countably Hilbertian topology θ in E for which both $X \in E'_\theta$ and $X' \in E'_\theta$ hold simultaneously a.s. $\square$

In the proof above we used Sazonov's idea [26] of using the inequality (6) and Yamazaki's idea [31] of applying integration with respect to the Gaussian measure to obtain (9) and (10). From Theorems 2.3.1 and 2.3.2 we can immediately deduce the following:

THEOREM 2.3.3 (the regularization theorem [16]). *An E-valued linear functional has a τ-regular version if and only if it is $I(\tau)$-continuous. Under this condition there exists a unique regular version.*

2.4. Characteristic functionals. Let (E, τ) be a multi-Hilbertian space.

DEFINITION 2.4.1. Let $X = \{X(f), f \in E\}$ be a linear random functional on E. The complex-valued function on E

$$C_X(f) := E(e^{iX(f)}), \qquad f \in E \tag{1}$$

is called the *characteristic functional* of X.

If X is a σ-concentrated E'_τ-valued random variable, then $\{X(f), f \in E\}$ is regarded as a linear random functional on E and so we can define the characteristic functional of X by (1).

If $X(f) = Y(f)$ a.s. for every $f \in E$, then $C_X = C_Y$. Hence, if X is a τ-regularization of a linear random functional Y, then $C_X = C_Y$.

The following relations between $\|X(f)\|_0$ and $C_X(f)$ for a linear random functional X are useful.

$$|C_X(f) - 1| \leq 2 \|X(f)\|_0, \tag{2}$$

$$\|X(f)\|_0^2 \leq c \int_0^1 |C_X(tf) - 1| \, dt \quad (c \text{ a positive constant}). \tag{3}$$

Observing that

$$|C_X(f) - 1| \leq E(|e^{iX(f)} - 1|)$$

and

$$|e^{ix} - 1| \leq |x| \wedge 2 \leq 2(|x| \wedge 1),$$

we obtain (2). Let us prove (3). Observing that

$$|C_X(tf) - 1| = |E(e^{itX(f)} - 1)| \geqq \operatorname{Re} E(1 - e^{itX(f)}) = E(1 - \cos tX(f)),$$

we obtain

$$\int_0^1 |C_X(tf) - 1|\, dt \geqq E\left(1 - \frac{\sin X(f)}{X(f)}, X(f) \neq 0\right)$$

$$\geqq c_1 E((|X(f)| \wedge 1)^2) \quad (c_1 \text{ a positive constant})$$

$$\geqq c_1 E(|X(f)| \wedge 1)^2,$$

because $f(x) = (1 - (\sin x)/x)/(|x| \wedge 1)^2$ $(0 < |x| < \infty)$ is positive, continuous and tends to $\frac{1}{6}$ (resp. 1) as $|x| \to 0$ (resp. ∞). Hence (3) holds for $c = 1/c_1$.

THEOREM 2.4.1. *Let X be a σ-concentrated E_τ'-valued random variable. Then $C(f) := C_X(f)$ satisfies the following:*

(C.1) $C(f)$ is positive definite:

$$\sum_{j,k=1}^n a_j \bar{a}_k C(f_j - f_k) \geqq 0, \qquad n \in \mathbf{N}, \quad a_j \in \mathbf{C}, \quad f_j \in E,$$

(C.2) $C(0) = 1$,

(C.3) $C(f)$ is $I(\tau)$-continuous at $f = 0$,

where $I(\tau)$ is the I-topology of τ introduced in the last section.

Conversely, if a complex-valued function $C(f)$ on E satisfies (C.1), (C.2) and (C.3), then we can construct a probability space (Ω, P) and a σ-concentrated E_τ'-valued random variable X on (Ω, P) such that $C = C_X$.

Proof. Suppose that X is a σ-concentrated E_τ'-valued random variable. Then $C = C_X$ obviously satisfies (C.1) and (C.2). Since X is σ-concentrated, the linear functional $\{X(f), f \in E\}$ is $I(\tau)$-continuous by Theorem 2.3.1. Hence for every $\varepsilon > 0$ we can find $p = p_\varepsilon < I(\tau)$ and $\delta = \delta_\varepsilon > 0$ such that $\|X(f)\|_0 < \varepsilon$ whenever $p(f) < \delta$. Hence

$$p(f) < \delta \Rightarrow \|X(f)\|_0 < \varepsilon \Rightarrow |C_X(f) - 1| < 2\varepsilon$$

by (2). Hence $C_X(f)$ is $I(\tau)$-continuous.

Suppose conversely that C satisfies (C.1), (C.2) and (C.3). For $f_1, f_2, \ldots, f_n \in E$ we define a function on $\mathbf{R}^n$;

$$C_{f_1 f_2 \cdots f_n}(t_1, t_2, \ldots, t_n) = C(t_1 f_1 + t_2 f_2 + \cdots + t_n f_n). \tag{4}$$

This function is positive definite and takes the value 1 at $0 \in \mathbf{R}^n$ by (C.1) and (C.2). For every $p < I(\tau)$ we have

$$p(t_1 f_1 + t_2 f_2 + \cdots + t_n f_n) \leqq \sum_k |t_k|\, p(f_k) \to 0$$

as $(t_1, t_2, \ldots, t_n) \to 0$ in $\mathbf{R}^n$. Hence $C_{f_1 f_2 \cdots f_n}$ is continuous at $0 \in \mathbf{R}^n$ by (4) and (C.3). Hence the classical Bochner theorem ensures that $C_{f_1 f_2 \cdots f_n}$ is the Fourier transform of a probability measure on $(\mathbf{R}^n, \mathscr{B}(\mathbf{R}^n))$, which we denote by $\mu_{f_1 f_2 \cdots f_n}$. Hence

$$\int_{\mathbf{R}^n} \exp\left(i \sum_{j=1}^n t_j x_j\right) \mu_{f_1 f_2 \cdots f_n}(dx) \quad (x = (x_1, x_2, \ldots, x_n))$$

$$= C(t_1 f_1 + t_2 f_2 + \cdots + t_n f_n) \tag{5}$$

by (4). Using this expression, we can easily check that the family $\{\mu_{f_1 f_2 \cdots f_n}, n \in \mathbf{N}, f_i \in E\}$ satisfies Kolmogorov's consistency condition. Hence we can construct a probability space (Ω, P) and a family of real random variables $Y(f)$, $f \in E$ such that the joint distribution of $Y(f_1), Y(f_2), \ldots, Y(f_n)$ is $\mu_{f_1 f_2 \cdots f_n}$. Then (5) implies that

$$E\left(\exp\left(i \sum_{j=1}^n t_j Y(f_j)\right)\right) = C(t_1 f_1 + t_2 f_2 + \cdots + t_n f_n). \tag{6}$$

$\{Y(f), f \in E\}$ is a linear random functional on E, namely

$$\sum_i t_i f_i = 0 \Rightarrow Z := \sum_i t_i Y(f_i) = 0 \text{ a.s.,}$$

because $E(e^{itZ}) = C(\sum_i tt_i f_i) = 1$, $t \in \mathbf{R}$ if $\sum_i t_i f_i = 0$. As a special case of (6) we have

$$C_Y(f) = E(e^{iY(f)}) = C(f),$$

i.e., C is the characteristic functional of a linear random functional of Y. By virtue of $(C.3)$ we can find $p = p_\varepsilon < I(\tau)$ and $\delta = \delta_\varepsilon > 0$ such that

$$p(f) < \delta \Rightarrow |C_Y(f) - 1| = |C(f) - 1| < \varepsilon.$$

If $p(f) < \delta$, then $p(tf) = |t| p(f) < \delta$ for $t \in [0, 1]$, and so

$$|C_Y(tf) - 1| < \varepsilon.$$

This implies that $\|Y(f)\|_0^2 \leq c\varepsilon$ by (3). Hence Y is an $I(\tau)$-continuous linear random functional on E, so Y has a τ-regular version X by Theorem 2.3.2. Hence $C_X(f) = C_Y(f) = C(f)$ for every $f \in E$. $\qquad \square$

Let E be a vector space, τ a multi-Hilbertian topology in E and μ a Lebesgue probability measure on $(E'_\tau, \mathscr{B}_K(E'_\tau))$ (see § 2.2). μ is called σ-*concentrated* if there exists a countably Hilbertian topology $\theta < \tau$ such that $\mu(E'_\theta) = 1$. Since E'_θ is a standard measurable space, we can use Theorem 2.2.2 to check that a σ-concentrated Lebesgue probability measure on E'_τ is perfect.

DEFINITION 2.4.2. Given a σ-concentrated Lebesgue probability measure μ on $E'_\tau = (E'_\tau, \mathscr{B}_K(E'))$,

$$C_\mu(f) := \int_{E_\tau'} e^{ix(f)} \mu(dx), \qquad f \in E \tag{7}$$

is called the *characteristic functional* of μ. A complex-valued function $C(f)$, $f \in E$ is called a *characteristic functional* on E if $C(f) = C_\mu(f)$ for some μ.

Note that the integral on the right-hand side of (7) is well defined because the map: $E'_\tau \to \mathbf{R} \ x \mapsto x(f)$ is measurable $\mathscr{B}_K(E'_\tau)/\mathscr{B}(\mathbf{R})$ by the definition of $\mathscr{B}_K(E'_\tau)$.

If X is a σ-concentrated E'_τ-valued random variable, then its probability distribution P^X is a σ-concentrated Lebesgue probability measure on $(E'_\tau, \mathscr{B}_K(E'_\tau))$, and $C_X = C_{P^X}$. If μ is a σ-concentrated Lebesgue probability measure on $(E'_\tau, \mathscr{B}_K(E'_\tau))$, then $(\Omega = E'_\tau, P = \mu)$ is a perfect probability space and

$$X(\omega) = \omega, \qquad \omega \in \Omega$$

is a σ-concentrated E'_τ-valued random variable such that $P^X = \mu$ and $C_X = C_\mu$.

THEOREM 2.4.2. *The correspondence $\mu \leftrightarrow C_\mu$ is $1-1$, namely if $C_\mu = C_\nu$, then* $\mu = \nu$.

Proof. For a finite set $F \subset E$ we denote the natural projection of E'_τ to $\mathbf{R}^F$ by $\pi_F : X \mapsto (X(f), f \in F)$ and the image measure $\pi_F\mu$ by μ_F. Then

$$C_\mu\left(\sum_{k=1}^n t_k f_k\right) = E\left(\exp\left(i \sum_{k=1}^n t_k Y(f_k)\right)\right) = \mathscr{F}\mu_F(t_1, t_2, \ldots, t_n)$$

$$(F = \{f_1, f_2, \ldots, f_n\}, \ \mathscr{F} : \text{Fourier transform}),$$

and similarly for C_ν. Hence $\mathscr{F}\mu_F = \mathscr{F}\nu_F$ and so $\mu_F = \nu_F$ by the classical Bochner theorem. Hence

$$\mu(B) = \nu(B) \quad \text{for } B = \pi_F^{-1}(A), \quad A \in \mathscr{B}(\mathbf{R}^n),$$

and so $\mu = \nu$ on $\mathscr{B}_K(E'_\tau)$ by the Dynkin class theorem. $\quad\square$

The following theorem due to A. N. Kolmogorov [17] includes the classical Bochner theorem ($E = \mathbf{R}^d$, $d < \infty$, $\tau = $ the Euclidean norm topology, $I(\tau) = \tau$), Sazonov's theorem [26] (E: separable Hilbert space, $\tau = $ the norm topology, $I(\tau) = $ the Sazonov (or Gross) topology) and Minlos' theorem [23] ($E = $ the rapidly decreasing functions $\mathscr{S}$, $\tau = $ the Schwartz topology, $I(\tau) = \tau$).

THEOREM 2.4.3 (the generalized Bochner theorem). *A complex-valued function $C(f)$, $f \in E$ is a characteristic functional if and only if $C(f)$ satisfies the conditions (C.1), (C.2) and (C.3) mentioned in Theorem 2.4.1.*

Proof. If $C = C_\mu$ for a σ-concentrated Lebesgue probability measure μ on $(E'_\tau, \mathscr{B}_K(E'_\tau))$, then $C = C_X$ for a σ-concentrated E'_τ-valued random variable X as we remarked after Definition 2.4.2. Hence $C(f)$ satisfies (C.1), (C.2) and (C.3) by the first part of Theorem 2.4.1.

Conversely, if C satisfies these three conditions, then $C = C_X$ for a σ-concentrated E'_τ-valued random variable X by the second part of Theorem 2.4.1. $\mu := P^X$ is a σ-concentrated Lebesgue measure on $(E'_\tau, \mathscr{B}_K(E'_\tau))$ and $C_X = C_\mu$, so $C = C_\mu$. $\quad\square$

The regularization theorem and the generalized Bochner theorem assert essentially the same fact from different angles, probabilistic and analytic respectively.

Remark 2.4.1. By examining the proof of Theorem 2.4.1, we can easily verify the following theorems on the characteristic functionals of linear random functionals.

THEOREM 2.4.4. *Let E be a vector space. If $Y: E \to L_0(\Omega, P)$ is a random linear functional on E, then $C(f) := C_Y(f)$ satisfies $(C.1)$, $(C.2)$ and*

$(C.3')$ $C(\sum_{j=1}^n t_j f_j)$ is continuous at $(t_1, t_2, \ldots, t_n) = (0, 0, \ldots, 0) \in \mathbf{R}^n$ for every $n \in \mathbf{N}$ and every $\{F_j\}$.

Conversely, if $C(f)$ satisfies $(C.1)$, $(C.2)$ and $(C.3')$, then we can construct a probability space (Ω, P) and a linear random functional $Y: E \to L_0(\Omega, P)$ such that $C = C_Y$.

THEOREM 2.4.5. *Let (E, α) be a topological vector space. If $Z: E \to L_0(\Omega, p)$ is an α-continuous linear random functional, then $C(f) := C_Z(f)$ satisfies $(C.1)$, $(C.2)$ and*

$(C.3'')$ $C(f)$ is α-continuous at $f = 0$.

Conversely if $C(f)$ satisfies $(C.1)$, $(C.2)$ and $(C.3'')$, then we can construct a probability space (Ω, P) and an α-continuous linear random functional $Z: E \to L_0(\Omega, P)$ such that $C = C_Z$.

2.5. Regular versions in the generalized sense. Let E be a vector space, α a multi-Hilbertian topology on E and $Y: E \to L_0(\Omega)$ an α-continuous linear random functional. Suppose that there exists a multi-Hilbertian topology τ on E such that

$$\alpha <_{HS} \tau. \tag{1}$$

Then $\alpha < I(\tau)$. Since Y is α-continuous, Y is obviously $I(\tau)$-continuous and so Y has a unique regular τ-version by Theorem 2.3.3. However, such a topology τ does not always exist. Therefore we introduce regular versions in the generalized sense.

DEFINITION 2.5.1. Let Y be an α-continuous linear random functional on E. If (i) there exists an α-dense vector subspace $E_1 \subset E$ and (ii) there exists a multi-Hilbertian τ_1 on E_1 such that the relative topology $\alpha|_{E_1}$ is HS-weaker than τ_1, then $Y_1 := Y|_{E_1}$ has a τ_1-regular version X, which is called a *regular version of Y in the generalized sense.*

Example 1. Wiener integral and white noise. Let $B(t)$, $0 \leq t \leq 1$ be a one-dimensional Wiener process. Then the *Wiener integral*

$$Y(f) := \int_0^1 f(t)\, dB(t), \qquad f \in L_2[0, 1]$$

is a $\|\cdot\|$-continuous linear random functional on $E := L_2[0, 1]$ $(\|\cdot\| = L_2$-norm on E). Let α be the $\|\cdot\|$-topology on E. There is no Hilbertian norm τ on E such that α is HS-weaker than τ. Let

$$E_1 := \mathscr{D}[0, 1], \qquad \|\varphi\|_{1/2} := \|\varphi'\| = \|D^{1/2}\varphi\|$$

where D is the operator introduced in § 1.4. Since the norm $\|\cdot\|$ restricted to E_1 is HS-weaker than $\|\cdot\|_{1/2}$, $Y|_{E_1}$ has a $\|\cdot\|_{1/2}$-regular version X, which is a regular version of Y in the generalized sense. X is a $\mathscr{D}'_{1/2}[0, 1]$-valued random variable represented as follows:

$$X(\omega)(\varphi) = -\int_0^1 \varphi'(t)B(t)\, dt = \partial B(\varphi), \qquad \varphi \in \mathscr{D}[0, 1],$$

where ∂ is the derivative in the distribution sense. X is called a *white noise* on $[0, 1]$.

Example 2. Gauss random measures and general white noises. Let m be a σ-finite measure on a measurable space $(S, \mathscr{B}(S))$ such that $L_2(S, m)$ is separable, and let

$$\mathscr{F} := \{F \in \mathscr{B}(S) : m(F) < \infty\}.$$

If $m : \mathscr{F} \to L_2(\Omega)$ satisfies the following conditions:

(i) $\{M(F)\}_F$ is a Gaussian system with mean 0,

(ii) $E(M(F_1)M(F_2)) = m(F_1 \cap F_2)$,

then $M = \{M(F)\}$ is called a *Gauss random measure* with *variance measure m*. These conditions imply that if $\{F_n\} \subset \mathscr{F}$ is a countable disjoint family, then $\{M(F_n)\}$ is independent and

$$M\left(\bigcup_n F_n\right) = \sum_n M(F_n) \quad \text{a.s.}$$

M induces a linear random functional on $E := L_2(S, \mu)$:

$$Y(f) = \int_S f(s)M(ds) \qquad \text{(Wiener integral)}.$$

Since $\|Y(f)\|_0 \leq \|Y(f)\|_2 = \|f\|$ (= the norm of f in $L_2(S, m)$), $Y(f)$ is $\|\cdot\|$-continuous in f. There is no Hilbertian norm on E that is HS-stronger than $\|\cdot\|$. We will find a regular version of Y in the generalized sense. Let $\{e_n\}$ be an ONB on $(E, \|\cdot\|)$ and E_1 the subspace of E that consists of all finite linear combinations of $\{e_n\}$. Then E_1 is dense in $(E, \|\cdot\|)$. Define a Hilbertian norm $\|\cdot\|_1$ on E_1 by

$$\|\varphi\|_1^2 := \sum_n n^2(\varphi, e_n)^2.$$

The right-hand side is in fact a finite sum, so $\|\cdot\|_1$ is well defined. $\{n^{-1}e_n\}$ is an ONB on $(E, \|\cdot\|_1)$ and

$$\sum_n \|n^{-1}e_n\|^2 = \sum_n n^{-2} < \infty.$$

Hence the restriction $Y|_{E_1}$ has a $\|\cdot\|_1$-regular version X, which is a regular version of Y in the generalized sense. X is an $E'_{1,\|\cdot\|_1}$-valued random variable and is called a *white noise* on S with *intensity measure m*. If S is $\mathbf{R}^n$ or a domain in $\mathbf{R}^n$, and if m is c-times the classical Lebesgue measure, then c is called the *intensity* of X. For example the white noise in Example 1 has intensity 1.

2.6. $\mathscr{D}'$-valued random variables. In this section we denote by $\mathscr{D}'$ one of the spaces introduced in §§ 1.4, 1.5 and 1.6 and use the following abbreviations:

$LR = LR(\mathscr{D}') :=$ the $\tau(\mathscr{D})$-continuous linear random functionals: $\mathscr{D} \to L_0(\Omega, P)$,

$RV = RV(\mathscr{D}') := $ the σ-concentrated $(\mathscr{D}', \mathscr{B}_K(\mathscr{D}'))$-valued random variables on (Ω, P),

$PM = PM(\mathscr{D}') := $ the σ-concentrated Lebesgue probability measures on $(\mathscr{D}', \mathscr{B}_K(\mathscr{D}'))$,

$CF = CF(\mathscr{D}') := $ the $\mathbf{C}$-valued functions defined on $\mathscr{D}$ that are positive definite, take the value 1 at $0 \in \mathscr{D}$ and are continuous at $0 \in \mathscr{D}$ with respect to the Schwartz topology $\tau(\mathscr{D})$.

Noting that $\tau = \tau(\mathscr{D})$ is nuclear, namely $I(\tau) = \tau$, and using the generalized Bochner theorem, we have a bijective map:

$$PM \to CF, \qquad \mu \mapsto C_\mu.$$

We also have the following maps:

(a) $LR \to RV$, $Y \mapsto$ the τ-regular version of Y,

(b) $RV \to PM$, $X \mapsto P^X$,

(c) $RV \to LR$, $X \mapsto \{X(f), f \in \mathscr{D}'\}$.

DEFINITION 2.6.1. A set $S \in \mathscr{B}_K(\mathscr{D}')$ is called a *measurable support* of $\mu \in PM$, if $\mu(S) = 1$. A measurable support of P^X is called a *measurable support* of X. A measurable support is simply called a *support* in the present discussion.

Let $\mathscr{S} := \mathscr{S}(\mathbf{R}^n)$ and $\tau(\mathscr{S})$ denote the Schwartz topology on $\mathscr{S}$, which is countably Hilbertian and nuclear. $\mathscr{D} = \mathscr{D}(\mathbf{R}^n)$ is a vector subspace of $\mathscr{S}$. $\mathscr{S}'$ is regarded as a vector subspace of $\mathscr{D}'$ by identifying $x \in \mathscr{S}'$ with $x\,|_{\mathscr{D}} \in \mathscr{D}'$. Then

$$\mathscr{S}' = \{x \in \mathscr{D}' : x(f) \text{ is } \tau(\mathscr{S})\,|_{\mathscr{D}}\text{-continuous in } f \in \mathscr{D}\}.$$

Since $\mathscr{B}_K(\mathscr{S}') = \mathscr{B}_K(\mathscr{D}') \cap \mathscr{S}'$, an $(\mathscr{S}', \mathscr{B}_K(\mathscr{S}'))$-valued random variable is regarded as an element of $RV(\mathscr{D}')$ of which $\mathscr{S}'$ is a support, and conversely.

THEOREM 2.6.1. *Suppose that* $p <_{HS} q < \tau \equiv \tau(\mathscr{D})$.

(i) *If* C_μ *is p-continuous, then*

$$\mathscr{D}'_q := \{x \in \mathscr{D}' : x(f) \text{ is q-continuous in } f\}$$

is a support of μ.

(ii) *If* $X \in RV(\mathscr{D}')$ *and if the linear random functional* $X : \mathscr{D} \to L_0(\Omega)$ *is p-continuous, then* $\mathscr{D}'_q$ *is a support of* X.

Proof. (i) The assumption $p <_{HS} q$ implies that $p < I(q)$. Being p-continuous, C_μ is $I(q)$-continuous. By the generalized Bochner theorem there exists a Lebesgue probability measure ν on $(\mathscr{D}'_q, \mathscr{B}_K(\mathscr{D}'_q))$ such that $C_\nu = C_\mu$. Observing that $\mathscr{D}'_q \in \mathscr{B}_K(\mathscr{D}')$, we can regard ν as an element of $PM(\mathscr{D}')$ that has a support $\mathscr{D}'_q$. But $C_\nu = C_\mu$ implies $\mu = \nu$. This completes the proof of (i).

(ii) Since $X : \mathscr{D} \to L_0(\Omega)$ is p-continuous, C_X is also p-continuous by

$$|E(e^{iX(f)} - 1)| \leq 2 \, \|X(f)\|_0 \qquad \text{(see (2) of § 2.4)}.$$

But $C_X = C_{P^X}$. Hence $\mathscr{D}'_q$ is a support of P^X by (i), so $\mathscr{D}'_q$ is a support of X. $\square$

DEFINITION 2.6.2. $X \in RV(\mathscr{D}')$ is said to be *Gauss distributed* if for every $f \in \mathscr{D}$ the real random variable $X(f)$ is Gauss distributed. Similarly for $X \in LR(\mathscr{D}')$.

THEOREM 2.6.2. $X \in RV(\mathscr{D}')$ *(or* $LR(\mathscr{D}')$*) is Gauss distributed if and only if* C_X *is* *represented in the form*

$$C_X(f) = \exp\{im(f) - \tfrac{1}{2}p(f)^2\}, \qquad f \in \mathscr{D} \tag{1}$$

where $m \in \mathscr{D}'$ *and* $p < \tau = \tau(\mathscr{D})$

Proof. If (1) holds, then

$$E(e^{itX(f)}) = C_X(tf) = \exp\{itm(f) - \tfrac{1}{2}t^2p(f)^2\}.$$

Hence $X(f)$ is Gauss distributed for every f, so X is Gauss distributed. This proves the *if* part.

Suppose that $X(f)$ is Gauss distributed, namely that

$$E(e^{itX(f)}) = \exp\left\{itm(f) - \frac{t^2}{2}V(f)\right\}, \tag{2}$$

where $m(f) = E(X(f))$ and $V(f) = V(X(f))$. $p(f) := V(f)^{1/2}$ is a seminorm on $\mathscr{D}$. p is Hilbertian, because it is derived from the symmetric positive definite bilinear form

$$p(f, g) = V(X(f), X(g)) \equiv \text{covariance } (X(f), X(g)).$$

$p(f)$ is τ-continuous at $f = 0$, because $C_X(f)$ is τ-continuous by $I(\tau) = \tau$ and

$$\exp\{-p(f)^2/2\} = |E(e^{iX(f)})| = |C_X(f)|.$$

Hence $p < \tau$.

$m(f)$ is also linear in f. We will prove that $m(f)$ is τ-continuous. Since both $C_X(f)$ and $V(f) = p(f)^2$ are τ-continuous at $f = 0$,

$$F(f) := \exp(im(f)) = C_X(f)\exp(V(f)/2)$$

is τ-continuous at $f = 0$. Hence we can find $q = q_\varepsilon < \tau$ and $\delta = \delta_\varepsilon$ such that

$$q(f) < \delta \Rightarrow |F(f) - 1| < \varepsilon. \tag{3}$$

Let N denote the standard Gaussian measure on $\mathbf{R}$,

$$\int_{\mathbf{R}} F(tf)N(dt) = \int_{\mathbf{R}} \exp(itm(f))N(dt) = e^{-m(f)^2/2}. \tag{4}$$

Take $a > 0$ so large that

$$\int_{|t| \geq a} N(dt) < \varepsilon.$$

If $q(f) < \delta/a$, then $q(tf) = |t|\,q(f) < \delta$ and so $|F(tf) - 1| < \varepsilon$ for $|t| < a$, which implies that

$$|e^{-m(f)^2} - 1| \leq \int_{|t| < a} |F(tf) - 1|\,N(dt) + \varepsilon < 2\varepsilon$$

by (4). This proves that $\exp\{-m(f)^2/2\}$ is τ-continuous at $f = 0$. Hence $|m(f)|$ is τ-continuous at $f = 0$. Since $m(f)$ is linear in f, $m(f)$ is τ-continuous, so we obtain $m \in \mathscr{D}'$, proving the *only if* part. $\square$

THEOREM 2.6.3. *Let $m \in \mathcal{D}'$ and $p < \tau$. Then there exists a unique Gaussian measure $\mu \in PM$ such that*

$$C_\mu(f) = \exp\{im(f) - \tfrac{1}{2}p(f)^2\}. \tag{5}$$

Proof. It suffices to check that the right-hand side of (5), written $C(f)$, satisfies the three conditions of the generalized Bochner theorem. (C.2) and (C.3) are obvious. To verify (C.1), observe that

$$C\left(\sum_{j=1}^n t_j f_j\right) = \exp\left\{i\sum t_j m(f_j) - \tfrac{1}{2}\sum_{jk} t_j t_k p(f_j, f_k)\right\}$$
$$= (\mathscr{F}N)(t_1 t_2 \cdots t_n),$$

where N is the n-dimensional Gaussian measure with mean vector $(m(t_j))_j$ and variance matrix $(p(f_j, f_k))_{j,k}$ and $\mathscr{F}$ denotes the Fourier transform. Hence $C(\sum t_j f_j)$ is positive definite in $(t_1, t_2, \ldots, t_n) \in \mathbf{R}^n$. This implies that $C(f)$ satisfies (C.1). $\square$

DEFINITION 2.6.3. The probability measure μ in the theorem above is called the *Gaussian measure* on $\mathcal{D}'$ with *mean (functional)* m and *variance (functional)* p^2, denoted by $N(m, p^2)$, in view of

$$m(f) = \int_{\mathcal{D}'} x(f)\mu(dx),$$

$$p(f)^2 = \int_{\mathcal{D}'} (x(f) - m(f))^2 \mu(dx).$$

$N(0, p^2)$ is called the *centered Gaussian measure* with variance (functional) p^2.

Suppose that μ is the Gaussian measure on $\mathcal{D}'$ with mean m and variance $V = p^2$. Since $p < \tau = I(\tau)$, there exists $q < \tau$ such that $p \prec_{HS} q$. Hence μ has a support $\mathcal{D}'_q$ by Theorem 2.6.1. Note that $\mathcal{D}'_q$ is a separable Hilbert space with the dual norm q'.

Suppose that X is Gauss distributed. Then P^X is a Gaussian measure. Hence X has a support $\mathcal{D}'_q$ for some $q < \tau$. Also $\{X(f), f \in \mathcal{D}\}$ is a Gaussian system with mean functional $m(f)$ and covariance functional $p(f, g)$ where m and p^2 are the mean and the variance of P^X respectively.

In the same way as in the Schwartz theory of distributions we define the *derivative* of a $\mathcal{D}'$-valued random variable $X(\omega)$ for each ω. For example, if $\mathcal{D} = \mathcal{D}(\mathbf{R})$, then

$$\partial X(f) := -X(f') \quad \text{for every } \omega.$$

∂X is obviously a $\mathcal{D}'$-valued random variable. Since X is σ-concentrated, we have a countably Hilbertian topology θ on $\mathcal{D}$ determined by $\{p_n\}$ such that $X \in \mathcal{D}'_\theta$ a.s. Let $\partial\theta$ denote the countably Hilbertian topology determined by $\{q_n(f) := p_n(f')\}$. Then $\partial X \in \mathcal{D}'_{\partial\theta}$ a.s. This proves that $\partial X \in RV$. For $\mathcal{D} = \mathcal{D}(\mathbf{R}^d)$ we can define ∂_i, $i = 1, 2, \ldots, d$ similarly. In case $\mathcal{D} = \mathcal{D}(M)$ we can define ∂_i locally, i.e., in each chart. Precisely speaking, if $(U, x = (x_1, x_2, \ldots, x_d))$ is a chart, then ∂_i can be defined for the restriction of X to $\mathcal{D}(U)$ and $\partial_i X\big|_{\mathcal{D}(U)} \in$

$RV(\mathscr{D}'(U))$. If a connection Γ is specified on M, we can globally discuss the derivative with respect to Γ.

Similarly we can define the *derivative* of a linear random functional. The derivative of a $\tau(\mathscr{D})$-continuous linear random functional is also $\tau(\mathscr{D})$-continuous.

A linear random functional $Y : \mathscr{D} \to L_0(\Omega)$ is called *Gauss distributed* if $Y(f)$ is Gauss distributed for every $f \in \mathscr{D}$. In this case every finite joint distribution of Y is Gaussian and the regular version of Y is also Gauss distributed.

The following theorem is obvious by the definition.

THEOREM 2.6.4. *The property of being Gauss distributed is inherited by derivatives for $\mathscr{D}'$-valued random variables and for linear random functionals on $\mathscr{D}$.*

Example 1. *Wiener processes.* Let $B = \{B(t, \omega)\}$ be a Wiener process on $(0, \infty)$ with $B(0+, \omega) = 0$. Since $B(\cdot, \omega) \in C = C(0, \infty)$, we can regard B as a $\mathscr{D}'$-valued random variable ($\mathscr{D} = \mathscr{D}(0, \infty)$):

$$B(f) = B(\omega)(f) = \int_0^\infty B(t, \omega) f(t)\, dt. \tag{6}$$

$B(f)$ is Gauss distributed with mean 0 and variance

$$V(f) = \int_0^\infty \int_0^\infty t \wedge s\, f(t) f(s)\, dt\, ds. \tag{7}$$

Since $p(f) = V(f)^{1/2}$ defines a Hilbertian norm and $p < \tau = \tau(\mathscr{D})$, we can find $q < \tau$ such that $p <_{HS} q$. Hence $B(\omega) \in \mathscr{D}'_q$ a.s., so $B \in RV = RV(\mathscr{D}')$.

Conversely, if $X \in RV$ is Gauss distributed with mean 0 and variance V defined by (7), then X is a Wiener process. To prove this we observe a linear random functional

$$X : \mathscr{D} \to L_0(\Omega), \qquad X(f)(\omega) = X(\omega)(f), \qquad f \in \mathscr{D}.$$

Observing that $X(\mathscr{D}) \subset L_2(\Omega)$ and that

$$\|X(f)\|_2 = p(f) \qquad (\|\cdot\|_2 = \text{the norm in } L_2(\Omega)),$$

we can extend $X : (\mathscr{D}, p) \to L_2(\Omega)$ to an isometric linear operator $\bar{X} : (\bar{\mathscr{D}}_p, \bar{p}) \to L_2(\Omega)$. Then X is a regular version of $\bar{X}$ in the generalized sense (§ 2.5). Noting that $\delta_t(f) \equiv f(t) \in \bar{\mathscr{D}}_p$, we define

$$\bar{B}(t) := \bar{X}(\delta_t).$$

Then $\{\bar{B}(t), t \in (0, \infty)\}$ is a Gaussian process with

$$E(\bar{B}(t)) = 0, \qquad E(\bar{B}(t)\bar{B}(s)) = \bar{p}(\delta_t, \delta_s) = t \wedge s,$$

so

$$E((\bar{B}(t) - \bar{B}(s))^4) = 3\, |t - s|^2.$$

Hence Kolmogorov's continuous version theorem ensures that $\bar{B}(t)$ has a continuous version, which we denote by $\{B(t)\}$. B is a Wiener process and induces a $\mathscr{D}'$-valued random variable by (6) and $\bar{B}$ induces a linear random functional:

$$\bar{B}(f) = \int_0^\infty B(t)f(t)\, dt \qquad \text{(Bochner integral in } L_2(\Omega)\text{)}.$$

Since $X(f) = \bar{X}(f) = \bar{B}(f) = B(f)$ a.s. for every $f \in \mathcal{D}$ and since X_ω, $B_\omega \in \mathcal{D}'$ a.s., we obtain $X_\omega = B_\omega$ a.s. This implies that X has a support $C = C(0, \infty)$.

Hence we can define a Wiener process B to be an element of $RV(\mathcal{D}')$ with characteristic functional:

$$\exp\,(-V(f)/2),$$

where $V(f)$ is defined by (7).

Example 2. Brownian sheets. Let $\mathcal{D} := \mathcal{D}((0, \infty)^2)$. $B \in RV(\mathcal{D}')$ is called a *Brownian sheet* on $(0, \infty)^2$, if

$$C_B(f) = \exp\,(-V(f)/2),$$

where

$$V(f) := \int_{(0,\infty)^4} (t_1 \wedge s_1)(t_2 \wedge s_2) f(t_1, t_2) f(s_1, s_2)\, dt_1\, dt_2\, ds_1\, ds_2.$$

B has a support $C((0, \infty)^2)$ (similarly for the d-dimensional case). This fact can be proved in the same way as in Example 1.

Example 3. White noise. Let B be a Brownian sheet on $(0, \infty)^d$. Then

$$W := \partial_1 \cdots \partial_d B \in RV(\mathcal{D}'), \qquad \mathcal{D} = \mathcal{D}((0, \infty)^d),$$

and

$$C_W(f) = \exp\,(-\|f\|^2/2) \tag{8}$$

where $\|\cdot\|$ is the norm in $L_2((0, \infty)^d)$. This $\mathcal{D}'$-valued random variable is called a *white noise* on $(0, \infty)^d$. A white noise has a support $\partial_1 \cdots \partial_d C$, where $C = C((0, \infty)^d)$.

We can directly define a white noise W to be an element of $RV(\mathcal{D}')$ whose characteristic functional is given by (8). To prove that W has a support $\partial_1 \cdots \partial_d C$ we consider a linear random functional $W : (\mathcal{D}, \|\cdot\|) \to L_2(\Omega)$ and its isometric linear extension $\bar{W}$. Defining

$$\bar{B}(t_1, \ldots, t_d) := \bar{W}(1_{(0,t_1) \times \cdots \times (0,t_d)})$$

where 1_A denotes the indicator of a set A, and using Totoki's extension of Kolmogorov's continuous version theorem to d dimensions (see Remark 2.6.1 at the end of this section), we can check that $\bar{B}$ has a continuous version, written B. Then B is a Brownian sheet and $W = \partial_1 \cdots \partial_d B$ a.s. This proves that W has a support $\partial_1 \cdots \partial_d C$.

Example 4. Gaussian processes. Let $\{Y(t), t \in (0, \pi)\}$ be a Gaussian process with mean 0 and covariance function $R(t, s) = E(Y(t)Y(s))$. Then Y induces a linear random functional on $\mathcal{D} \equiv \mathcal{D}(0, \pi)$:

$$Y(f) = \int_0^\pi Y(t)f(t)\, dt \qquad \text{(Bochner integral)},$$

if $R(t, s)$ is bounded and continuous on $(0, \pi)^2$. (This is true in a more general situation.) Then $Y(f)$ is obviously continuous and

$$C_Y(f) = \exp\{-V(f)/2\},$$

where

$$V(f) = \iint R(t, s)f(t)f(s)\, dt\, ds, \qquad f \in \mathcal{D}.$$

Since

$$\|Y(f)\|_0^2 \leq \|Y(f)\|_2^2 = V(f) \leq c\,\|f\|^2,$$

where c is a positive constant and $\|\cdot\|$ is the norm in $L_2(0, \pi)$, $Y(f)$ is $\tau(\mathcal{D})$-continuous. Hence Y has a regular version $\in RV(\mathcal{D}')$, denoted by the same notation Y.

Suppose that $R \in C^\infty((0, \pi)^2)$. Then

$$R(t, s) = \sum_{m,n=1}^\infty a_{mn} s_m(t) s_n(s),$$

where $\{s_n\}$ was introduced in § 1.4 and

$$a_{mn} = o((mn)^{-2k-1}) \quad \text{for every } k.$$

Then

$$V(f) \leq \sum a_{mn}(f, s_m)(f, s_n) \leq c\left(\sum_n n^{-2k-1}|(f, s_n)|\right)^2 \leq c'\sum_n n^{-4k}(f, s_n)^2 = c'\,\|f\|_{-k}^2,$$

namely, $\|Y(f)\|_0 \leq c'\,\|f\|_{-k}$, where c and c' are positive constants. Since $\|\cdot\|_{-k} <_{HS} \|\cdot\|_{-k+1}$, we can use Theorem 2.6.1 to obtain

$$Y \in \mathcal{D}'_{-k+1} \quad \text{a.s. for every } k.$$

Hence $Y \in \bigcap_k \mathcal{D}'_{-k+1} \subset C^\infty$ a.s.

Since the property of being C^∞ is local, we can use this method to discuss the cases $\mathcal{D} = \mathcal{D}(M)$ and $\mathcal{D}(p, q)(M)$ by observing Y in each chart.

Remark 2.6.1. The following theorem is a version of Totoki's extension of Kolmogorov's continuous version theorem. Kolmogorov's theorem is the case $d = 1$.

THEOREM 2.6.5. *Let* $\{X_t, t \in t\}$ *(T: closed and convex in $\mathbf{R}^d$) be a family of random variables with values in a complete metric space (S, d). If the following condition holds for every n:*

$$E[d(X_s, X_t)^{\alpha}n] \leq \beta_n\,|s - t|^{d+\gamma_n},$$

or slightly more generally

$$P\{d(X_s, X_t) > \delta\} \leq \beta_n \delta^{-\alpha_n}\,|s - t|^{d+\gamma_n} \quad \text{whenever } \delta > 0,$$

for every s, $t \in T \cap [-n, n]^d$, where α_n, β_n, γ_n are positive constants depending only on n, then X_t has a continuous version.

See [4] for the proof.

2.7. $\mathcal{D}'$-valued stochastic processes. Hereafter a σ-concentrated $\mathcal{D}'$-valued random variable is called a $\mathcal{D}'$-*variable*. A family of $\mathcal{D}'$-variables indexed by a time parameter ranging over an interval T is called a $\mathcal{D}'$-*valued stochastic process* or a $\mathcal{D}'$-*process*. Here we assume that $T = [0, \infty)$.

Let

$$X_t(\omega) = X_t(\omega, f) \quad (f \in \mathcal{D}), \qquad t \in T$$

be a $\mathcal{D}'$-process. If we fix ω, we obtain a $\mathcal{D}'$-valued function of $t \in T$, called the *sample process* of the $\mathcal{D}'$-process $\{X_t, t \in T\}$ *corresponding* to ω. The sample process is denoted by $X_{\cdot}(\omega)$. It is obvious that $X_{\cdot}(\omega) \in (\mathcal{D}')^T$. The map

$$X_{\cdot} : \Omega \to (\mathcal{D}')^T, \qquad \omega \mapsto X_{\cdot}(\omega)$$

is measurable $\mathcal{D}(P)/\mathcal{D}_K((\mathcal{D}')^T)$, as we can see from the fact that for every $t \in T$ X_t is a $\mathcal{D}'$-variable, namely $\omega \mapsto X_t(\omega)$ is measurable $\mathcal{D}(P)/\mathcal{D}_K(\mathcal{D}')$.

A stochastic process $\{X_t\}$ is said to be *continuous i.p.* (*i.p.* = in probability) if for every $t_0 \in T$, every $\varepsilon > 0$ and every $\tau_s(\mathcal{D}')$-neighborhood U of $0 \in \mathcal{D}'$ there exists $\delta = \delta(t_0, \varepsilon, U)$ such that

$$t \in T, \ |t - t_0| < \delta \Rightarrow P\{X_t - X_{t_0} \in U\} > 1 - \varepsilon,$$

where $\tau_s(\mathcal{D}')$ is the strong topology in $\mathcal{D}'$ defined in § 1.2.

$\{X_t\}$ is said to be *sample continuous* (resp. *sample continuous a.s.*) if for every (resp. almost every) ω the sample process

$$X_{\cdot}(\omega) : T \to \mathcal{D}', \qquad t \mapsto X_t(\omega)$$

is $\tau_s(\mathcal{D}')$-continuous. Here we can replace $\tau_s(\mathcal{D}')$ by $\tau_w(\mathcal{D}')$, because $(\mathcal{D}', \tau_s(\mathcal{D}'))$ is a Montel space [30].

Let $\{X_t\}$ and $\{Y_t\}$ be $\mathcal{D}'$-processes. If

$$Y_t = X_t \quad \text{a.s. for every } t$$

(the exceptional ω-set may depend on t), then $\{Y_t\}$ is said to be *equivalent to* $\{X_t\}$. If

$$Y_{\cdot} = X_{\cdot} \quad \text{a.s.,}$$

namely if

$$P(Y_t = X_t \text{ for every } t) = 1,$$

then $\{Y_t\}$ is said to be *strictly equivalent* to $\{X_t\}$. Strict equivalence implies equivalence but not conversely.

If $\{X_t\}$ is continuous i.p., then every $\mathcal{D}'$-process equivalent to $\{X_t\}$ is continuous i.p. In other words continuity in probability is inherited by equivalent processes. Similarly sample continuity is inherited by strictly equivalent processes, but not by equivalent ones.

A sample continuous $\mathcal{D}'$-process equivalent to a $\mathcal{D}'$-process $\{X_t\}$ is called a *continuous version* of $\{X_t\}$. All continuous versions of a $\mathcal{D}'$-process are strictly equivalent to each other.

A $\mathcal{D}'$-process $\{X_t\}$ is said to *have independent increments* if for every $\{t_i\}$ $(0 \leq t_0 < t_1 < \cdots < t_n)$ the increments $X_{t_i} - X_{t_{i-1}}$, $i = 1, 2, \ldots, n$ are independent. Furthermore if the probability distribution of the increment $X_t - X_s$ $(s < t)$ is invariant under time shift, then $\{X_t\}$ is said to *have stationary independent increments*.

If $\{X_t\}$ has independent increments, then for every $f \in \mathcal{D}$, $\{X_t(f)\}_t$ is a real-valued process with independent increments, but not conversely. From the definition of independence we can easily see that $\{X_t\}$ has independent increments if and only if for every n and every $\{f_1, f_2, \ldots, f_n\} \subset \mathcal{D}$ the n-dimensional stochastic process

$$(X_t(f_1), X_t(f_2), \ldots, X_t(f_n)), \qquad t \in T$$

has independent increments.

A $\mathcal{D}'$-process $\{X_t\}$ is called *Gaussian* if the family of real random variables $X_t(f)$, $t \in T$, $f \in \mathcal{D}$ is a Gaussian system. Furthermore if $X_t(f)$ has mean 0 for every (t, f), then $\{X_t\}$ is called *centered Gaussian*.

Generalizing one-dimensional Wiener processes, we introduce *Wiener $\mathcal{D}'$-processes*.

DEFINITION 2.7.1. A sample continuous process $\{B_t\}$ with stationary independent increments is called a Wiener $\mathcal{D}'$-process, if $B_0 = 0$.

Let $\{B_t\}$ be a Wiener $\mathcal{D}'$-process. Then for every $f \in \mathcal{D}$, $\{B_t(f)\}_t$ is a sample continuous real-valued process with stationary independent increments. Hence $\{B_t(f)\}_t$ is a Gaussian process with

$$E(B_t(f)) = m(f) \cdot t \quad \text{and} \quad V(B_t(f)) = V(f) \cdot t,$$

where $m(f) \in \mathbf{R}$ and $V(f) \geq 0$, so the $\mathcal{D}'$-variable B_t is Gauss distributed. Hence

$$m \in \mathcal{D}' \quad \text{and} \quad V = p^2 \qquad (p < \tau(\mathcal{D}))$$

by Theorem 2.6.2. If $t > s$, then

$$V(B_t(f), B_s(g)) = V(B_t(f) - B_s(f), B_s(g)) + V(B_s(f), B_s(g)) = p(f, g)s$$

because of the independence of $B_t - B_s$ and B_s. Hence

$$V(B_t(f), B_s(g)) = p(f, g)(s \wedge t), \qquad f, g \in \mathcal{D}, \quad s, t \in T.$$

Thus we obtain the following theorem.

THEOREM 2.7.1. *If $\{B_t\}$ is a Wiener $\mathcal{D}'$-process, then $\{B_t(f)\}_{t,f}$ is a Gaussian system with*
  (i) *mean functional $E(B_t(f)) = m(f) \cdot t$,*
  (ii) *variance functional $V(B_t(f), B_s(g)) = p(f, g)(t \wedge s)$,*
where $m \in \mathcal{D}'$ and $p < \tau(\mathcal{D})$.

Now we will prove the existence theorem for Wiener $\mathcal{D}'$-processes.

THEOREM 2.7.2. *Given $m \in \mathcal{D}'$ and $p < \tau = \tau(\mathcal{D})$, there exists a Wiener $\mathcal{D}'$-process $\{B_t\}$ such that*

$$E(B_1(f)) = m(f) \quad and \quad V(B_1(f)) = p(f)^2. \tag{1}$$

Such a process is unique in law in the following sense: if $\{B_t\}$ and $\{B'_t\}$ are such processes, then

$$P^{B.} = P^{B'.} \quad i.e. \quad P(B. \in \Gamma) = P(B'. \in \Gamma), \quad \Gamma \in \mathcal{B}_K((\mathcal{D}')^T).$$

Proof of existence. First we consider the case $m = 0$. Let

$$V((t, f), (s, g)) = p(f, g)(t \wedge s).$$

Then $V((t, f), (s, g))$ is symmetric and positive definite in (t, f) and (s, g). Symmetry is obvious. Observing

$$p(f, g)(t \wedge s) = \int_T p(e_t(u)f, e_s(u)g)\, du \qquad (e_t(u) := 1_{[0,t]}(u)),$$

we obtain

$$\sum_{i,j=1}^{n} a_i a_j V((t_i, f_i), (t_j, f_j)) = \int_T p\left(\sum_i a_i e_{t_i}(u)f\right)^2 du \geqq 0$$

$$(n \in \mathbf{N}, a_i \in \mathbf{R}, f_i \in \mathcal{D}),$$

proving that $V((t, f), (s, g))$ is positive definite. Hence there exists a centered Gaussian system $\{B_t(f)\}_{t,f}$ such that

$$E(B_t(f)B_s(f)) = V((t, f), (s, g)) = p(f, g)(t \wedge s).$$

Fix t for the moment. Then

$$B_t : \mathcal{D} \to L_0(\Omega), \qquad f \mapsto B_t(f)$$

is linear and τ-continuous. If $\sum_i a_i f_i = 0$, then

$$E\left(\left(\sum_i a_i B_t(f_i)\right)^2\right) = \sum_{ij} a_i a_j p(f_i, f_j)t = tp\left(\sum_i a_i f_i\right)^2 = 0,$$

so $\sum a_i B_t(f_i) = 0$, which proves that B_t is linear. Observing that

$$\|B_t(f)\|_0^2 \leqq \|B_t(f)\|_2^2 = E(B_t(f)^2) = tp(f)^2,$$

we can see that $B_t(f)$ is p-continuous and so τ-continuous in f. The regularization theorem (Theorem 2.3.3) ensures that B_t has a τ-regular version, which we denote with the same notation B_t.

Now we will prove that the $\mathcal{D}'$-process $\{B_t\}$ has a continuous version. Since $p < \tau = I(\tau)$, we can find $q < \tau$ such that $p <_{HS} q$. Then the $\mathcal{D}'$-variable B_t has a support $\mathcal{D}'_q$ by Theorem 2.6.1. $\mathcal{D}'_q$ is a separable Hilbert space with dual norm q'. By modifying B_t on a P-null ω-set for each t, we can assume that $B_t \in \mathcal{D}'_q$ for every $\omega \in \Omega$.

Observing that $B_t - B_s$ is centered Gaussian and that

$$E((B_t - B_s)(f)^2) = p(f)^2(t - 2t \wedge s + s) = p(f)^2 |t - s|,$$

we obtain

$$E(q'(B_t - B_s)^4) \leqq 3 |t - s|^2 (p:q)^4_{HS}, \tag{2}$$

setting $E = \mathcal{D}$, $X = (t-s)^{-1/2}(B_t - B_s)$ and $\alpha = 2$ in the next theorem. Using Kolmogorov's continuous version theorem extended to processes with values in a complete separable metric space, we can find a q'-continuous version of $\{B_t\}$, denoted by $\{B_t\}$ again. Since $q < \tau$ and since the q'-topology in $\mathcal{D}'_q$ coincides with the relative topology $\tau_s(\mathcal{D})|_{\mathcal{D}'_q}$, the version obtained above is obviously a sample continuous $\mathcal{D}'$-process and satisfies all the conditions mentioned in the theorem in case $m = 0$. Then $\{mt + B_t\}_t$ satisfies all the conditions for the general case.

The proof of uniqueness is routine. Since $\{B_t\}$ and $\{B'_t\}$ are sample continuous and have stationary independent increments, $\{B'_t(f)\}_{t,f}$ and $\{B_t(f)\}_{t,f}$ are Gaussian systems with the same mean functional and variance functional. Hence we obtain

$$P\{B'_{t_i}(f_i) < a_i, i = 1, 2, \ldots, n\} = P\{B'_{t_i}(f_i) < a_i, i = 1, 2, \ldots, n\}.$$

Use the Dynkin class theorem to complete the proof. $\square$

THEOREM 2.7.3. *Let p and q be separable H-seminorms on a vector space E such that $p <_{HS} q$ and let q' be the dual norm of q on the dual space E'_q. If $X(\omega)$ is a centered Gaussian (E'_q, q')-valued random variable with $E(X(f)^2) = p(f)^2$, then*

$$E(q'(x)^{2\alpha}) \leqq c_\alpha \lambda^{2\alpha} \qquad (\alpha \geqq 1, \text{ equality holds for } \alpha = 1)$$

where c_α is the 2αth order moment of the standard Gauss measure on $\mathbf{R}$ and $\lambda = (p:q)_{HS} < \infty$. Hence $E(q'(x)^{2\alpha}) \leqq c_\alpha E(q'(x)^2)^\alpha$.

Proof. Let $\{e_n, n = 1, 2, \ldots\}$ be a q-ONB on E. Then

$$\lambda^2 = \sum_n p(e_n)^2.$$

Let $\{e'_n\}$ be the q'-ONB on E'_q dual to $\{e_n\}$, i.e.

$$e'_m(e_n) = \delta_{mn} \qquad (m, n = 1, 2, \ldots).$$

Since $X(\omega) \in E'_q$, we have an orthogonal series expansion:

$$X(\omega) = \sum_n X(e_n)e'_n \qquad (X(e_n) = X(\omega)(e_n)), \tag{3}$$

and so

$$E(q'(X)^2) = E\left(\sum_n X(e_n)^2\right) = \sum_n p(e_n)^2 = \lambda^2.$$

This proves the case $\alpha = 1$.

To prove the case $\alpha > 1$ we will use a nice trick due to D. Stroock. Let $\{a_n(t)\}$ be the Rademacher ONB on $L_2[0, 1]$. Because of the Gaussian property of X, $X(e_n)$, $n = 1, 2, \ldots$ are independent centered Gaussian variables, each $X(e_n)$ having variance $p(e_n)^2$, and so are $X(e_n)a_n(t)$, $n = 1, 2, \ldots$ for every t, because $a_n(t) = \pm 1$. Hence

$$S(t) := \sum_{n=1}^{\infty} a_n(t)X(e_n) \tag{4}$$

is a centered Gaussian variable with variance λ^2, so

$$E(S(t)^{2\alpha}) = c_\alpha \lambda^{2\alpha}, \qquad 0 \leq t \leq 1. \tag{5}$$

Since $\{a_n(t)\}$ and $\{e_n'\}$ are ONB's on $L_2[0,1]$ and E_q' respectively, (3) and (4) imply that

$$\int_0^1 S(t)^2 \, dt = \sum_{n=1}^{\infty} X(e_n)^2 = q'(X)^2, \tag{6}$$

and so

$$\int_0^1 S(t)^{2\alpha} \, dt \geq \left(\int_0^1 S(t)^2 \, dt \right)^{\alpha} = q'(X)^{2\alpha} \qquad (\alpha \geq 1).$$

This, combined with (5), implies that

$$E(q'(X)^{2\alpha}) \leq \int_0^1 E(S(t)^{2\alpha}) \, dt = c_\alpha \lambda^{2\alpha}. \qquad \square$$

In view of Theorems 2.7.1 and 2.7.2, a Wiener $\mathscr{D}'$-process $\{B_t\}$ with $E(B_1(f)) = m(f)$ and $V(B_1(f)) = p(f)^2$ is called a *Wiener $\mathscr{D}'$-process* (m, p^2). If $\mathscr{D} = \mathscr{D}(R^n)$, $m = 0$ and $p =$ the $L_2(\mathbf{R}^n)$-norm, then it is called a *standard Wiener $\mathscr{D}'$-process*.

Let $D = D(T, \mathscr{D}')$ denote the vector space of all functions $T \to \mathscr{D}'$ that are right $\tau_s(\mathscr{D}')$-continuous and have left $\tau_s(\mathscr{D}')$-limits at all points of T. We can replace $\tau_s(\mathscr{D}')$ by $\tau_w(\mathscr{D}')$ because $\mathscr{D}'$ is a Montel space, so we will say right (or left) continuous.

A $\mathscr{D}'$-process $\{X_t\}$ is called a *D-sample $\mathscr{D}'$-process* if every sample process of $\{X_t\}$ belongs to D. A D-sample $\mathscr{D}'$-process $\{X_t\}$ with independent increments is called a *Lévy $\mathscr{D}'$-process* if it is continuous in probability and $X_0 = 0$. Furthermore if the probability distribution of the increment $X_t - X_s$ $(t > s)$ is invariant under time shift, then $\{X_t\}$ is called a *time homogeneous Lévy $\mathscr{D}'$-process*. A Wiener $\mathscr{D}'$-process is a special case of a time homogeneous Lévy $\mathscr{D}'$-process.

Let $\{X_t\}$ be a time homogeneous Lévy $\mathscr{D}'$-process. Then the sample process $X_.$ is a $(D, \mathscr{B}_K(D))$-valued random variable. Although the proof is omitted here, $(D, \mathscr{B}_K(D))$ is a standard measurable space, so the probability distribution of $X_.$, $P^{X_.}$ is a Lebesgue probability measure on $(D, \mathscr{B}_K(D))$. Because of the property of stationary independent increments, $P^{X_.}$ is completely determined by giving

$$C_{X_t}(f) \equiv e(e^{iX_t(f)}) \qquad (\text{or } C_{X_1}(f)).$$

For example, the distribution of the sample process of a Wiener process (m, p) $\{B_t\}$ is characterized by

$$E(e^{iB_t(f)}) = \exp\{itm(f) - tp(f)^2/2\}.$$

A time homogeneous Lévy $\mathscr{D}'$-process $\{X_t\}$ is called a *symmetric stable $\mathscr{D}'$-process*, if

$$E(e^{iX_t(f)}) = \exp\{-tp(f)^{2\beta}\} \qquad (p < \tau = \tau(\mathscr{D}), 0 < \beta \leq 1).$$

To prove the existence of this process, first recall the following idea due to S. Bochner.

Let $\{B_t\}$ be a one-dimensional Wiener process and $\{T_t = T_t^\beta\}$ $(0 < \beta \leqq 1)$ an increasing stable process such that

$$E(e^{-\alpha T_t^\beta}) = \exp\{-t(2\alpha)^\beta\} \qquad (\alpha \geqq 0). \tag{7}$$

Suppose that $\{B_t\}$ and $\{T_t\}$ are independent. Then

$$X_t := B_{T_t} \qquad (0 \leqq \beta \leqq 1) \tag{8}$$

is a symmetric stable process with

$$E(e^{izX_t}) = \exp\{-t\,|z|^{2\beta}\} \qquad (z \in \mathbf{R}).$$

If we replace the one-dimensional Wiener process $\{B(t)\}$ by the Wiener $\mathscr{D}'$-process $(0, p)$, then we can easily check that $X_t := B_{T_t}$ is a time homogeneous Lévy $\mathscr{D}'$-process, which is characterized by

$$\begin{aligned}
E(e^{iX_t(f)}) &= E(e^{iB_{T_t}(f)}) \\
&= \int_0^\infty E(e^{iB_s(f)})P^{T_t}(ds) \\
&= \int_0^\infty e^{-sp(f)^2/2}P^{T_t}(ds) = E(\exp(-T_t p(f)^2/2)) \\
&= \exp\{-tp(f)^{2\beta}\} \quad \text{by (7),}
\end{aligned}$$

as we wanted.

If $p < \tau(\mathscr{S})$, then P^{X_t} is concentrated on $\mathscr{S}'$, so we can regard $\{X_t\}$ as a symmetric stable $\mathscr{S}'$-process. See X. Fernique [5] for a discussion of $\mathscr{S}'$-processes.

2.8. Linear random operators. For any $p < \tau = \tau(\mathscr{D})$ the Hilbertian seminormed space $(\mathscr{D}, p)$ yields a separable Hilbert space $(\bar{\mathscr{D}}_p, \bar{p})$ by completion (§ 1.1), and the dual space of $(\mathscr{D}, p)$, $(\mathscr{D}'_p, p')$ is regarded as the dual space of $(\bar{\mathscr{D}}_p, \bar{p})$. $(\mathscr{D}'_p, p')$ is isomorphic to $(\bar{\mathscr{D}}_p, \bar{p})$ under the bijection

$$\alpha_p : \bar{\mathscr{D}}_p \to \mathscr{D}'_p, \qquad f \mapsto \bar{p}(f, \cdot), \tag{1}$$

so that

$$\xi(f) = \bar{p}(\alpha_p^{-1}\xi, f) = p'(\xi, \alpha_p f), \qquad \xi \in \mathscr{D}'_p, \quad f \in \bar{\mathscr{D}}_p. \tag{2}$$

If $\{e_n\}$ is an ONB on $(\bar{\mathscr{D}}_p, \bar{p})$, then $\{\alpha_p e_n\}$ is its dual ONB on $(\mathscr{D}'_p, p')$.

Let $_1\mathscr{D}$ and $\mathscr{D}$ be two spaces of the types in §§ 1.4, 1.5 and 1.6; for example, $_1\mathscr{D} = \mathscr{D}(\mathbf{R}^d)$ and $\mathscr{D} = \mathscr{D}(m, n)(M)$. Let

$\mathscr{L}_{p,u} := $ the bounded linear operators: $(_1\mathscr{D}'_p, p') \to (\mathscr{D}'_u, u')$,

$\mathscr{L}_{u,p} := $ the bounded linear operators: $(\mathscr{D}'_u, u') \to (_1\mathscr{D}'_p, p')$,

$\tilde{\mathscr{L}}_{u,p} := $ the bounded linear operators: $(\bar{\mathscr{D}}_u, \bar{u}) \to (_1\bar{\mathscr{D}}_p, \bar{p})$.

The conjugate operator of $L \in \mathscr{L}_{p,u}$ is denoted by L^*. Define $\tilde{L} := \alpha_p^{-1}L^*\alpha_u$.

Then

$$u'(L\xi, \eta) = p'(\xi, L^*\eta), \qquad \xi \in {}_1\mathscr{D}'_p, \quad \eta \in \mathscr{D}'_u, \tag{3}$$

$$L^* \in \mathscr{L}_{u,p}, \qquad \tilde{L} \in \tilde{\mathscr{L}}_{u,p}. \tag{4}$$

$\tilde{L}$ is also a bounded linear operator, and

$$(L\xi)(f) = \xi(\tilde{L}f), \qquad \xi \in {}_1\mathscr{D}'_p, \quad f \in \bar{\mathscr{D}}_u \tag{5}$$

because we can use (2) to check that

$$(L\xi)(f) = u'(L\xi, \alpha_u f) = p'(\xi, L^*\alpha_u f) = p'(\xi, \alpha_p\tilde{L}f) = \xi(\tilde{L}f).$$

The operator norm and the *HS*-norm are denoted by $\|\cdot\|$ and $\|\cdot\|_{HS}$ respectively. If $L \in \mathscr{L}_{p,u}$, then $\|L\| < \infty$ and $\|L\|_{HS} \leq \infty$. It is known that $\|L\| = \|L^*\|$. Since both α_p and α_u are isomorphic mappings, $\|\tilde{L}\| = \|L^*\|$. Similarly for the *HS*-norms $\|L\|_{HS}$, $\|L^*\|_{HS}$ and $\|\tilde{L}\|_{HS}$. Hence

$$\|L\| = \|L^*\| = \|\tilde{L}\|, \qquad \|L\|_{HS} = \|L^*\|_{HS} = \|\tilde{L}\|_{HS}. \tag{6}$$

Let $\mathscr{B}_K(\mathscr{L}_{p,u})$ be the Kolmogorov σ-algebra on $\mathscr{L}_{p,u}$, the σ-algebra generated by the sets

$$\{L \in \mathscr{L}_{p,u} : (L\xi)(f) < a\}, \qquad \xi \in {}_1\mathscr{D}'_p, \quad f \in \bar{\mathscr{D}}_p, \quad a \in \mathbf{R}.$$

An $\mathscr{L}_{p,u}$-valued function $L = L_\omega$ on (Ω, P) is called an $\mathscr{L}_{p,u}$-valued random variable (abbr. $\mathscr{L}_{p,u}$-variable) if it is measurable $\mathscr{D}(P)/\mathscr{B}_K(\mathscr{L}_{p,u})$. If L_ω is an $\mathscr{L}_{p,u}$-variable, then L_ω^* is an $\mathscr{L}_{u,p}$-variable and $\tilde{L}_\omega$ is an $\tilde{\mathscr{L}}_{u,p}$-variable.

Suppose that $X = X_\omega$ is a linear random functional on ${}_1\mathscr{D}$ such that

$$E(X(f)^2) \leq cp(f)^2, \qquad f \in {}_1\mathscr{D} \quad (c > 0), \tag{7}$$

and that $L = L_\omega$ is an $\mathscr{L}_{p,u}$-variable. X can be regarded as a ${}_1\mathscr{D}'$-variable by taking a regular version. In view of (5) we define a u-continuous linear random functional on $\mathscr{D}$, LX by

$$(LX)(f) := X(\tilde{L}f) \tag{8}$$

for each $f \in \mathscr{D}$, if the right-hand side has any meaning.

First consider the case where L is deterministic, namely L does not depend on ω. For each $f \in \mathscr{D} \subset \bar{\mathscr{D}}_u$ we have $\tilde{L}f \in {}_1\bar{\mathscr{D}}_p$. The map $f \mapsto X(f)$ is a bounded linear operator: $({}_1\mathscr{D}, p) \to L_2(\Omega)$ by the assumption (7), so it can be extended to a unique bounded linear operator: $({}_1\bar{\mathscr{D}}_p, \bar{p}) \to L_2(\Omega)$. Since $\tilde{L}f \in {}_1\bar{\mathscr{D}}_p$, $X(\tilde{L}f)$ is meaningful, so $LX(f)$ is well defined. $LX : \mathscr{D} \to L_2(\Omega)$ is obviously linear. It is also u-continuous, because

$$E((LX)(f)^2) \leq c\bar{p}(\tilde{L}f)^2 \leq c\|\tilde{L}\|^2 u(f)^2 = c\|L\|^2 u(f)^2. \tag{9}$$

PROPOSITION 2.8.1. *Let $\{Y(f)\}$ be a u-continuous linear random functional on $\mathscr{D}$ such that $E(Y(f)^2) \leq cu(f)^2$ and $\{d_n\}$ an ONB on $(\mathscr{D}, u)$. If*

$$\sum_n E(Y(d_n)^2) < \infty,$$

then Y has a u-regular version, for which

$$E(u'(Y)^2) = \sum_n E(Y(d_n)^2). \tag{10}$$

Proof. Define $r(f) \geqq 0$ by

$$r(f)^2 = E(Y(f)^2).$$

Then r is an H-seminorm $< u$. By the assumption Y is an r-continuous linear random functional and

$$\sum_n r(d_n)^2 < \infty \quad \text{and so} \quad r <_{HS} u.$$

Now use the regularization theorem. $\quad\square$

Applying this proposition to LX, we can prove that if $\|L\|_{HS} < \infty$, then LX has a u-regular version, because for any ONB $\{d_n\}$ on $(\mathcal{D}, u)$ we obtain

$$\sum_n E((LX)(d_n)^2) \leqq c \sum_n p(\tilde{L} d_n)^2 = c \|\tilde{L}\|_{HS}^2 = c \|L\|_{HS}^2 < \infty.$$

Now we will consider the case where L is random.

THEOREM 2.8.1. *Suppose that* (a) $L = L_\omega$ *is an* $\mathscr{L}_{p,u}$*-variable,* (b) X *satisfies* (7) *and* (c) L *and* X *are independent.*

(i) *If* $E(\|L\|^2) < \infty$, *then* $\{(LX)(f)\}$ *is a linear random functional and*

$$E((LX)(f)^2) \leqq cE(\|L\|^2)u(f)^2.$$

(ii) *Furthermore, if* $E(\|L\|_{HS}^2) < \infty$, *then* LX *is a* $\mathscr{D}_u'$*-variable (and so a* $\mathscr{D}'$*-variable) by taking a u-regular version, and*

$$E(u'(LX)^2) \leqq cE(\|L\|_{HS}^2). \qquad (Equality \ holds \ if \ E(X(f)^2) = cp(f)^2.)$$

Proof. Since X and L are independent, the conditional probability measure $P(X \in \cdot \mid L)$ coincides with $P(X \in \cdot)$.

$$E(X(f)^2 \mid L) = E(X(f)^2) \leqq cp(f)^2 \quad \text{a.s.} \quad \text{by (7)}.$$

Since L is deterministic on $(\Omega, P(\cdot \mid L))$, $(LX)(f)$ is well defined a.s. $(P(\cdot \mid L))$. Hence $(LX)(f)$ is well defined a.s. (P). Using (9), we obtain

$$E((LX)(f)^2 \mid L) \leqq cp(\tilde{L}f)^2 \leqq c \|L\|^2 u(f)^2 \quad \text{a.s.},$$

or

$$E((LX)(f)^2) \leqq cE(\|L\|^2)u(f)^2.$$

Hence, if $E(\|L\|^2) < \infty$, then $\{(LX)(f)\}$ is a u-continuous linear random functional. If $E(\|L\|_{HS}^2) < \infty$ then we have, for any ONB $\{d_n\}$ on $(\mathcal{D}, u)$,

$$\sum_n E((LX)(d_n)^2) \leqq \sum_n cE(\bar{p}(\tilde{L} d_n)^2) = cE(\|\tilde{L}\|_{HS}^2) = cE(\|L\|_{HS}^2) < \infty.$$

Now use Proposition 2.8.1 to complete the proof. $\quad\square$

The following representation of LX may be useful.

THEOREM 2.8.2. *Suppose that* (a) $L = L_\omega$ *is an* $\mathscr{L}_{p,u}$*-variable,* (b') $X = X_\omega$ *satisfies* $E(X(f)^2) = cp(f)^2$ *and* (c) *they are independent. Let* $\{e_n\}$ *be an ONB on* $(_1\mathscr{D}, p)$ *and* $\{\varepsilon_n\}$ *its dual ONB on* $(_1\mathscr{D}'_p, p')$.

(i) *If* $E(\|L\|^2) < \infty$, *then*

$$(LX)(f) = \sum_n X(e_n)(L\varepsilon_n)(f) \quad a.s., \quad f \in \bar{\mathscr{D}}_u \tag{11}$$

where the infinite sum is convergent in $\|\cdot\|_2$ *for each f. (Remember that* $\|\cdot\|_2$ *and* $(\cdot, \cdot)_2$ *are the norm and the inner product in* $L_2(\Omega)$ *respectively.)*

(ii) *If* $E(\|L\|^2_{HS}) < \infty$, *then*

$$LX = \sum_n X(e_n)L\varepsilon_n \quad a.s. \quad and \quad E(u'(LX)^2) = cE(\|L\|^2_{HS}), \tag{12}$$

where the infinite sum is u'*-convergent in* $\|\cdot\|_2$.

Proof. (i) Fixing any $f \in \bar{\mathscr{D}}_u$, we obtain

$$(LX)(f) = X(\tilde{L}f) \quad \text{a.s.} \quad \text{(by the definition)}$$
$$= p'(X, \alpha_p \tilde{L}f)$$
$$= \sum_n p'(X, \varepsilon_n)p'(\alpha_p \tilde{L}f, \varepsilon_n)$$
$$= \sum_n X(e_n)(L\varepsilon_n)(f) \quad \text{(by } \varepsilon_n = \alpha_p e_n).$$

This proves (11), where the infinite sum is convergent a.s.

Denoting $X(e_n)L\varepsilon_n$ by Y_n and observing that $\{X(e_n)\}_n$ and $\{(L\varepsilon_n)(f)\}_n$ are independent, we obtain

$$E(Y_n(f)^2) = E(X(e_n)^2(L\varepsilon_n)(f)^2)$$
$$= E(X(e_n)^2)E((L\varepsilon_n)(f)^2)$$
$$= cp(e_n)^2 E((L\varepsilon_n)(f)^2)$$
$$\leq cE(\|L\|^2)u(f)^2 < \infty.$$

Hence $Y_n(f) \in L_2(\Omega)$. Similarly

$$E(Y_n(f)Y_m(f)) = cp(e_n, e_m)E((L\varepsilon_n)(f)(L\varepsilon_m)(f))$$
$$= 0 \quad (n \neq m).$$

Hence $\{Y_n(f)\}_n$ is an orthogonal sequence in $L_2(\Omega)$. But

$$c^{-1}\sum_n \|Y_n(f)\|^2_2 = \sum_n E((L\varepsilon_n)(f)^2) = E\left(\sum_n \varepsilon_n(\tilde{L}f)^2\right)$$
$$= E(\bar{u}(\tilde{L}f)^2) \leq E(\|L\|^2)u(f)^2$$
$$= E(\|L\|^2)u(f)^2 < \infty.$$

Hence $\sum_n Y_n(f)$ is convergent in $\|\cdot\|_2$.

(ii) Since $E(\|L\|^2_{HS}) < \infty$ by the assumption, we obtain $E(\|L\|^2) < \infty$, so (11) holds by (i). Hence, for the proof of (ii) it suffices to check that $\sum_n Y_n$ is u'-convergent in $\|\cdot\|_2$.

Denote by $L_2(\Omega, \mathcal{D}_u')$ the $\mathcal{D}_u'$-variables Y such that

$$\|Y\|_{2,u'}^2 := E(u'(Y)^2) < \infty.$$

Then $L_2(\Omega, \mathcal{D}_u')$ is a Hilbert space with norm $\|\cdot\|_{2,u'}$ and inner product

$$(Y, Z)_{2,u'} := E(u'(Y, Z)).$$

Now observe that

$$\begin{aligned}
E(u'(Y_n)^2) &= E(u'(X(e_n)L\varepsilon_n)^2) \\
&= E(X(e_n)^2 u'(L\varepsilon_n)^2) \\
&= E(X(e_n)^2)E(u'(L\varepsilon_n)^2) \\
&\leq cE(\|L\|^2) < \infty.
\end{aligned}$$

Hence $Y_n \in L_2(\Omega, \mathcal{D}_u')$. Similarly

$$\begin{aligned}
(Y_n, Y_m)_{2,u'} &= E(u'(Y_n, Y_m)) \\
&= E(X(e_n)X(e_m))E(u'(L\varepsilon_n, L\varepsilon_m)) \\
&= cp(e_n, e_m)E(u'(L\varepsilon_n, L\varepsilon_m));
\end{aligned}$$

this expression vanishes for $n \neq m$ and equals $cE(u'(L\varepsilon_n)^2)$ for $n = m$. Hence $\{Y_n\}$ is an orthogonal sequence in $L_2(\Omega, \mathcal{D}_u)$ and

$$\sum_n \|Y_n\|_{2,u'}^2 = c\sum_n E(u'(L\varepsilon_n)^2) = cE(\|L\|_{HS}^2) < \infty.$$

Hence $\sum Y_n$ is convergent in $\|\cdot\|_{2,u'}$, namely $\sum Y_n$ is u'-convergent in $\|\cdot\|_2$, and the second formula of (12) follows. $\quad\square$

Since $L\xi$ is bilinear in the pair (L, ξ), we may expect that $L_\omega X_\omega$ is bilinear in (L_ω, X_ω). We will discuss this problem.

Let $\mathbf{L}_p = \mathbf{L}_p(_1\mathcal{D})$ be the linear random functionals X's on $_1\mathcal{D}$ such that

$$E(X(f)^2) \leq cp(f)^2. \tag{13}$$

The infimum of the square roots of all such c's is called the *p-norm* of X, denoted by $\|X\|_p$. Similarly we define $\mathbf{L}_u = \mathbf{L}_u(\mathcal{D})$ and the *u*-norm of its element. Also we define the *HS p-norm* $|X|_p \geq 0$ for $X = X_\omega \in \mathbf{L}_p$ by

$$|X|_p^2 = \sum_n E(X(e_n)^2), \tag{14}$$

where $\{e_n\}$ is an ONB in $(_1\mathcal{D}, p)$; this norm is well defined independently of the choice of the ONB. Similarly we define $|Y|_u^2$ for $Y = Y_\omega \in \mathbf{L}_u(\mathcal{D})$.

$X \in \mathbf{L}_p$ yields a unique linear random functional $\bar{X}$ on $(\bar{\mathcal{D}}_p, \bar{p})$ whose *p*-norm equals $\|X\|_p$. We denote $\bar{X}$ by the same notation X if there is no possibility of confusion. We can choose an ONB on $(_1\bar{\mathcal{D}}, \bar{p})$ to define the *HS p-norm* of X. If the *HS p-norm* of X is finite, then X yields an $_1\mathcal{D}_p'$-variable by taking a *p*-regular version $\sum_n X(e_n)\varepsilon_n$, denoted by X again, if there is no possibility of confusion, where $\{\varepsilon_n\}$ is the dual norm of $\{e_n\}$ and similarly for $Y \in \mathbf{L}_u$.

For $L \in \mathcal{L}_{p,u}$ we define its (p, u)-norm $\|L\|_{p,u}$ and its *HS* (p, u)-norm $|L|_{p,u}$ in the usual way, i.e.

$$\|L\|_{p,u} = \sup \{u'(L\xi) : \xi \in {}_1\mathcal{D}'_p, \, p'(\xi) \leqq 1\}, \tag{15}$$

$$|L|^2_{p,u} = \sum_n u'(Le_n)^2, \tag{16}$$

where $\{e_n\}$ is an ONB in ${}_1\mathcal{D}_p$ and the *HS* (p, u)-norm is well defined independently of the ONB. In the discussion above $\|L\|_{p,u}$ and $|L|_{p,u}$ were denoted by $\|L\|$ and $\|\cdot\|_{HS}$. Note that $\|L\|_{p,u}$ and $|L|_{p,u}$ depend on ω.

Let $\mathbf{L}_{p,u} = \mathbf{L}_{p,u}({}_1\mathcal{D}'_p, \mathcal{D}'_u)$ be the $\mathcal{L}_{p,u}$-variables L's such that $E(\|L\|^2_{p,u}) < \infty$. In case $X_\omega \in \mathbf{L}_p$ and $L_\omega \in \mathbf{L}_{p,u}$ are independent we defined $L_\omega X_\omega \in \mathbf{L}_u$ in the discussion above as follows:

$$L_\omega X_\omega(f) := X_\omega(\tilde{L}_\omega f), \tag{17}$$

where $\tilde{L} = \alpha_p^{-1} L \alpha_u$.

Let $\mathcal{F}$ be a σ-subalgebra of $\mathcal{D}(P)$. $X_\omega \in \mathbf{L}_p$ is called $\mathcal{F}$-*measurable* if for every f the map $\omega \mapsto X_\omega(f)$ is measurable $\mathcal{F}/\mathcal{B}(\mathbf{R})$. Similarly $L_\omega \in \mathbf{L}_{p,u}$ is called $\mathcal{F}$-*measurable* if for every $\xi \in {}_1\mathcal{D}'_p$ and every $g \in \mathcal{D}_u$ the map $\omega \mapsto (L_\omega \xi)(g)$ is measurable $\mathcal{F}/\mathcal{B}(\mathbf{R})$.

$\mathbf{L}_p$ is a vector space with the usual linear operation. We denote by $\mathbf{L}_{p,\mathcal{F}} = \mathbf{L}_{p,\mathcal{F}}({}_1\mathcal{D})$ the $\mathcal{F}$-measurable elements of $\mathbf{L}_p$. $\mathbf{L}_{p,\mathcal{F}}$ is a vector subspace of $\mathbf{L}_p$. Similarly we define $\mathbf{L}_{u,\mathcal{F}} = \mathbf{L}_{u,\mathcal{F}}(\mathcal{D})$ and $\mathbf{L}_{p,u,\mathcal{F}} = \mathbf{L}_{p,u,\mathcal{F}}({}_1\mathcal{D}'_p, \mathcal{D}'_u)$.

Suppose that $\mathcal{F}_1$ and $\mathcal{F}_2$ are independent σ-subalgebras of $\mathcal{D}(P)$ and that $\mathcal{F}$ is the union σ-algebra of $\mathcal{F}_1$ and $\mathcal{F}_2$. If every $L_\omega \in \mathbf{L}_{p,u,\mathcal{F}_1}$ and every $X_\omega \in \mathbf{L}_{p,\mathcal{F}_2}$ are independent, then $L_\omega X_\omega$ is well defined by Theorem 2.8.1. and belongs to $\mathbf{L}_{u,\mathcal{F}}$. The following theorem follows easily from the definitions and Theorem 2.8.1.

THEOREM 2.8.3. *If $\mathcal{F}_1$ and $\mathcal{F}_2$ are independent, then $L_\omega X_\omega \in \mathbf{L}_{p,\mathcal{F}}$ is bilinear in $L_\omega \in \mathbf{L}_{p,u,\mathcal{F}_1}$ and $X_\omega \in \mathbf{L}_{p,\mathcal{F}_2}$, and*

$$\|L_\omega X_\omega\|^2_u \leqq E(\|L_\omega\|^2_{p,u})\|X_\omega\|^2_p, \tag{18}$$

$$|L_\omega X_\omega|^2_u \leqq E(|L_\omega|^2_{p,u}) \|X_\omega\|^2_p. \tag{19}$$

Both sides of (18) *and* (19) *may be* ∞.

Remark 2.8.1. Suppose that ${}_1\mathcal{D} = \mathcal{D}$ and $p <_{HS} u$. Then the inclusion map $I_{p,u} : \mathcal{D}'_p \to \mathcal{D}'_u$ belongs to $\mathcal{L}_{p,u} \subset \mathbf{L}_{p,u}$. Since $I_{p,u}$ is deterministic, it is independent of all $X_\omega \in \mathbf{L}_p$. Hence

$$I_{p,u}X = \text{the } u\text{-regular version of } X.$$

If X is a $\mathcal{D}'_q$-variable and $E(X(f)^2) \leqq cp(f)^2$, then X, viewed as a linear random functional, belongs to $\mathbf{L}_p$ and X, viewed as a $\mathcal{D}'_q$-variable, is the q-regular version of a linear random functional X and so coincides with the u-regular version of X a.s. Hence

$$X = I_{p,u}X \quad \text{a.s.}$$

In the discussion above we fixed p and u. Now we fix p and move u. If $u < v$, then

$$\mathscr{D}'_u \subset \mathscr{D}'_v, \quad \mathscr{L}_{p,u} \subset \mathscr{L}_{p,v}, \quad \mathbf{L}_{p,u} \subset \mathbf{L}_{p,v}.$$

Let $\mathscr{L}_{p,\cdot} := \bigcup_u \mathscr{L}_{p,u}$ and $\mathbf{L}_{p,\cdot} := \bigcup_u \mathbf{L}_{p,u}$. If $L \in \mathbf{L}_{p,\cdot}$ and $X \in \mathbf{L}_{p,\cdot}$ are independent, then LX is well defined and is a $\tau(\mathscr{D})$-continuous random linear functional, and (18) and (19) hold for every $u < \tau(\mathscr{D})$. The facts mentioned above hold for the inclusion map $I_p : \mathscr{D}'_p \to \mathscr{D}'$ with the obvious modification.

Remark 2.8.2. If $X \in \mathbf{L}_p(_1\mathscr{D})$, then there exists $q < \tau(_1\mathscr{D})$ such that $p <_{HS} q$, so X has a q-regular version which equals $I_{p,q} X_\omega$ as we explained above. Suppose that $M \in \mathbf{L}_{q,u}$. Then $I_{p,q} X_\omega \in \mathscr{D}'_q$ and $M_\omega : \mathscr{D}'_q \to \mathscr{D}'_u$ for every ω, so we can apply M_ω to $I_{p,q} X_\omega$ sample-wise to obtain $M_\omega(I_{p,q} X_\omega) \in \mathscr{D}'_u$. But $MI_{p,q} \in \mathbf{L}_{p,u}$ and

$$|MI_{p,q}|_{p,u} \leq \|M\|_{q,u} \, |I_{p,q}|_{p,q},$$

so

$$E(|MI_{p,q}|^2_{p,u}) \leq E(\|M\|^2_{p,u}) |I_{p,q}|^2 < \infty;$$

note that $|I_{p,q}|_{p,q} = (p : q)_{HS}$. Hence we can use Theorem 2.8.1(ii) to see that $(MI_{p,q})X$ is a $\mathscr{D}'_u$-variable. It is easy to check that

$$M(I_{p,q}X) = (MI_{p,q})X.$$

Remark 2.8.3. Suppose that $E(X(f)^2) \leq cp(f)^2$ and $p <_{HS} q$. Then X has a version $\in \mathscr{D}'_q$. Let Y be a $\mathscr{D}'$-variable represented by a function $\{Y(x, \omega)\}_x$ such that $Y(\cdot)f(\cdot) \in \bar{\mathscr{D}}_q$ a.s. for every $f \in \mathscr{D}$. Recalling that X is regarded as a bounded linear functional on $\bar{\mathscr{D}}_q$ for every ω, we define a $\mathscr{D}'_q$-variable YX by

$$(YX)(f) = X(Yf).$$

If Y and X are independent, then

$$E[(YX)(f)^2] \leq cE(p(Yf)^2).$$

CHAPTER 3

Infinite Dimensional Stochastic Differential Equations

3.1. General remarks. A typical example of stochastic differential equations in finite dimensions takes the form

$$dX_t = \sigma(X_t)\, dB_t + b(X_t)\, dt, \qquad t \in T(=[0, \infty)), \tag{1}$$

namely

$$X_t = X_0 + \int_0^t \sigma(X_s)\, dB_s + \int_0^t b(X_s)\, ds, \qquad t \in T, \tag{2}$$

where

   (i) $\{B_t\}$ is a given Wiener $\mathbf{R}^r$-process,

   (ii) $\{X_t\}$ is a sample continuous $\mathbf{R}^d$-process that is *adapted* to $\{\mathscr{F}_t\}$, where

$$\mathscr{F}_t = \mathscr{F}_t(B) := \text{the } \sigma\text{-algebra generated by } B_s, s \leq t, t \in T.$$

$\{X_t\}$ is said to be *adapted* to $\{\mathscr{F}_t\}$ if for every $t \in T$, X_t is measurable $\mathscr{F}_t/\mathscr{B}(\mathbf{R}^d)$. The adaptedness assumption is imposed to guarantee that the stochastic integral (the first integral of the right-hand side of (2)) is meaningful.

   (iii) $\sigma = \sigma(x)\ (x \in \mathbf{R}^d)$ is a linear operator from $\mathbf{R}^r$ into $\mathbf{R}^d$, i.e. a $d \times r$-matrix.

   (iv) $b = b(x)\ (x \in \mathbf{R}^d)$ is a d-dimensional vector.

Under the Lipschitz condition

$$\|\sigma(x) - \sigma(y)\| + \|b(x) - b(y)\| \leq c\, \|x - y\| \tag{3}$$

we can prove that (1) (or (2)) has a unique solution with any given initial condition. This classical theory has recently been extensively developed [21], [6], [29], [12], [4], [10].

In the theory of infinite dimensional stochastic differential equations both the given Wiener process $\{B_t\}$ and the solution process $\{X_t\}$ take values in infinite dimensional spaces. A number of papers, lecture notes and books have been published on infinite dimensional stochastic integrals and stochastic differential equations; see H. H. Kuo [18], M. Metivier [22], Ja. I. Belopol'skaya and Ja. D. Daleckiĭ [1], etc. for general theory and P. Malliavin [20], D. W. Stroock [28], B. Gaveau [8], D. A. Dawson [3], T. Funaki [7], etc. for special problems. However, the results are formulated in different frameworks in accordance with the different purposes of the authors. Hence it is not easy to compare one discussion with another and to apply general results to special problems.

In our discussion we restrict ourselves to the case where the given Wiener process $\{B_t\}$ is a Wiener ${}_1\mathscr{D}'$-process $(0, p^2)$ (§ 2.7) and the solution process $\{X_t\}$ is a $\mathscr{D}'$-process, where ${}_1\mathscr{D}'$ and $\mathscr{D}'$ are Schwartz spaces of distributions (§§ 1.4, 1.5 and 1.6). For example ${}_1\mathscr{D}' = \mathscr{D}'(\mathbf{R}^d)$ and $\mathscr{D}' = \mathscr{D}(m, n)'(M)$, although ${}_1\mathscr{D}' = \mathscr{D}'$

in most special problems. Even under this restriction we can cover practically all infinite dimensional stochastic differential equations appearing in applications.

In this section we briefly explain stochastic integrals and stochastic differential equations in our framework. In the subsequent sections of the chapter we will deal with some special equations from our viewpoint.

(A). Stochastic integrals. Let $\{B_t\}$ be a Wiener ${}_1\mathscr{D}'$-process $(0, p^2)$ and $\mathscr{L} = \mathscr{L}(p)$ the continuous linear maps of ${}_1\mathscr{D}'_p$ into $\mathscr{D}'$. $\mathscr{L}$ is a measurable space with the Kolmogorov σ-algebra $\mathscr{B}_K(\mathscr{L})$, the σ-algebra generated by the sets

$$\{L \in \mathscr{L} : (L\xi)(f) < a\}, \qquad \xi \in {}_1\mathscr{D}'_p, \quad f \in \mathscr{D}', \quad a \in \mathbf{R}.$$

We will define the stochastic integral

$$I_t(\sigma) = \int_0^t \sigma_s \, dB_s, \qquad t \in T = [0, \infty), \tag{4}$$

where $\{\sigma_t\}$ is an $(\mathscr{L}, \mathscr{B}_K(\mathscr{L}))$-valued stochastic process (abbr. $\mathscr{L}$-process) satisfying the conditions explained below.

For any $u < \tau(\mathscr{D}')$ we introduced the space of operators $\mathscr{L}_u = \mathscr{L}_{p,u}$ in § 2.8. $\mathscr{B}_K(\mathscr{L}_u)$ is the restriction of $\mathscr{B}_K(\mathscr{L})$ to $\mathscr{L}_u$, namely

$$\mathscr{B}_K(\mathscr{L}_u) = \mathscr{B}_K(\mathscr{L}) \cap \mathscr{L}_u.$$

For $L \in \mathscr{L}$ we define the operator norm $\|L\|_u = \|L\|_{p,u}$ and the HS-norm $|L|_u = |L|_{p,u}$ as follows: If $L \in \mathscr{L}_u$, then

$$\|L\|_u = \sup \{|u'(L\xi)|, \, p'(\xi) \leq 1\},$$

$$|L|_u = \text{the } HS\text{-norm of } L : ({}_1\mathscr{D}'_p, p') \to (\mathscr{D}'_u, u'),$$

as we defined in § 2.8. If $L \notin \mathscr{L}_u$, then these norms are defined to be ∞.

We denote by $\mathscr{F}_t = \mathscr{F}_t(B)$ the σ-algebras on Ω generated by $\{B_s(f), s \leq t, f \in {}_1\mathscr{D}\}$ and the P-null sets. We impose the following assumptions on $\{\sigma_t\}_t$.

$(\sigma.1)$ *(adaptability).* For every $(t, \xi, f) \in T \times {}_1\mathscr{D}' \times \mathscr{D}$ the map: $[0, t] \times \Omega \to \mathbf{R}$, $(s, \omega) \mapsto (\sigma_s(\omega)\xi)(f)$ is measurable $\mathscr{B}[0, t] \times \mathscr{F}_t / \mathscr{B}(\mathbf{R})$.

$(\sigma.2)$ *(local integrability).* There exist a sequence of H-seminorms $u_n < \tau(\mathscr{D})$, $n = 1, 2, \ldots$ and an increasing sequence of $\mathscr{F}_\cdot$-stopping times $\theta_n = \theta_n(\omega) \uparrow \infty$ a.s. such that

$$E\left(\int_0^\infty |\sigma_{s \wedge \theta_n}|_{u_n}^2 \, ds\right) < \infty, \tag{5}$$

where $|\cdot|_u = |\cdot|_{p,u}$.

First we fix t and observe the case where $\sigma_s \in \mathscr{L}_u$ and $E(|\sigma_s|_u^2) < \infty$ for every $s \in [0, t)$ and every ω and there exists a subdivision of $[0, t) : 0 = s_0 < s_1 < \cdots < s_n = t$ such that

$$\sigma_s = \sigma_{s_{i-1}}, \qquad s \in [s_{i-1}, s_i), \quad i = 1, 2, \ldots, n.$$

In this case we define

$$I_t(\sigma) := \sum_{i=1}^{n} \sigma_{s_{i-1}} (B_{s_i} - B_{s_{i-1}}).$$ (8)

Each summand of the right-hand side defines a $\mathcal{D}'_u$-variable. Denoting $\sigma_{s_{i-1}}$, $B_{s_i} - B_{s_{i-1}}$ and $s_i - s_{i-1}$ by σ_i, B_i and Δ_i respectively, we obtain

$$E[u'(\sigma_i B_i)^2] = E(|\sigma_i|_u^2)\, \Delta_i$$

by Theorem 2.8.1(ii), because $E(B_i(f)^2) = \Delta_i p(f)^2$, σ_i and B_i are independent by the adaptability assumption and $E(|\sigma_i|_u^2) < \infty$.

For every $f \in \bar{\mathcal{D}}_u$ we have

$$E(\sigma_i B_i(f)^2) \leqq E(u'(\sigma_i B_i)^2) u(f)^2 = E(|\sigma_i|_u^2)\, \Delta_i u(f)^2 < \infty.$$

Hence, if $i < j$, then

$$E(|\sigma_i B_i(f)\sigma_j B_j(f)|) < \infty,$$

so

$$\begin{aligned}
E(\sigma_i B_i(f)\sigma_j B_j(f)) &= E(\sigma_i B_i(f) B_j(\tilde{\sigma}_j f)) \\
&= E(\sigma_i B_i(f) E(B_j(g))|_{g=\tilde{\sigma}_j f}) \quad \text{by adaptability} \\
&= 0.
\end{aligned}$$

Hence

$$E(I_t(\sigma)(f)^2) = \sum_{i=1}^{n} E((\sigma_i B_i(f))^2).$$

Let $\{d_k\}$ be an ONB on $(\bar{\mathcal{D}}_u, \bar{u})$. Setting $f = d_k$ in the equality above and summing up in k, we obtain

$$E(u'(I_t(\sigma))^2) = \sum_{i=1}^{n} E(u'(\sigma_i B_i)^2) = \sum_{i=1}^{n} E(|\sigma_i|_u^2)\, \Delta_i,$$

namely

$$E(u'(I_t(\sigma))^2) = E\left(\int_0^t |\sigma_s|_u^2\, ds\right).$$ (7)

Once this is done, we can define $I_t(\sigma)$ for the general case exactly in the same way as for the one-dimensional stochastic integral, and we obtain a sample continuous version of $\{I_t(\sigma)\}_t$. Also we can verify that $\{I_t(\sigma)\}$ has all the fundamental properties that hold in the one-dimensional case.

(B). Stochastic differential equations. Now we are in a position to discuss the following *stochastic differential equation*:

$$dX_t = \sigma(X_t)\, dB_t + b(X_t)\, dt, \qquad X_0 = \xi,$$ (11)

namely

$$X_t = \xi + \int_0^t \sigma(X_s)\, dB_s + \int_0^t b(X_s)\, ds,$$ (12)

where $\sigma : \mathcal{D}' \to \mathcal{L}$ is measurable $\mathcal{B}_K(\mathcal{D}')/\mathcal{B}_K(\mathcal{L})$ and $b : \mathcal{D}' \to \mathcal{D}'$ is measurable $\mathcal{B}_K(\mathcal{D}')/\mathcal{B}_K(\mathcal{D}')$. The following types of problems are most common in applications.

Type 1. To find a sample r'-continuous $\mathcal{D}'_r$-process $\{X_t\}$ satisfying (11) in case

$$\sigma(\mathcal{D}'_r) \subset \mathcal{L}_r \quad \text{and} \quad b(\mathcal{D}'_r) \subset \mathcal{D}'_r,$$

where $r < \tau(\mathcal{D})$.

Type 2. To find a sample $\tau_s(\mathcal{D}')$-continuous $\mathcal{D}'$-process $\{X_t\}$ such that (i) $X_t(\omega) \in \mathcal{D}'_r$ a.s. for each t and (ii) $\{X_t\}$ satisfies (11) in case

$$\sigma(\mathcal{D}'_r) \subset \mathcal{L}_u \quad \text{and} \quad b(\mathcal{D}'_r) \subset \mathcal{D}'_u,$$

where $r < u < \tau(\mathcal{D})$.

Type 1 is a special case ($r = u$) of Type 2. For Type 1 the Lipschitz condition

$$\|\sigma(x) - \sigma(y)\|_r + r'(b(x) - b(y)) \leqq Kr'(x - y) \tag{13}$$

is sufficient for the existence and the uniqueness of the solution, as we can verify by successive approximations similar to those for the finite dimensional case.

For equations of Type 2 no general method is known. We have to use a special device for each problem; see §§ 3.4 and 3.5.

A simple example of Type 1 is the Ornstein–Uhlenbeck equation with friction constant $\frac{1}{2}$:

$$dX_t = dB_t - \tfrac{1}{2}X_t\, dt, \tag{14}$$

where $_1\mathcal{D} = \mathcal{D} = \mathcal{D}(\mathbf{R})$, $p = \|\cdot\|_0$, $u = r = \|\cdot\|_1$ and dB_t is interpreted as $I_{p,r}\, dB_t$; see § 1.3 for $\|\cdot\|_0$ and $\|\cdot\|_2$ and Remark 2.8.2 for $I_{p,r}$. The solution of this equation with $X_0 = \xi$ is unique and is given by

$$X_t = e^{-t/2}\xi + \int_0^t e^{-(t-s)/2}\, dB_s, \tag{15}$$

where dB is interpreted as $I_{p,r}\, dB_s$ as above. We can easily see that there is no other solution even if we admit solutions taking values in $\mathcal{D}' \backslash \mathcal{D}'_r$. (See the next section.)

A simple example of Type 2 is the following equation:

$$dX_t = dB_t + \tfrac{1}{2}DX_t\, dt, \tag{16}$$

where $_1\mathcal{D} = \mathcal{D} = \mathcal{D}(\mathbf{R})$, $p \equiv \|\cdot\|_0$, $u = \|\cdot\|_1$ and $r = \|\cdot\|_0$ and where $D : \mathcal{D}'_0(\mathbf{R}) \equiv \mathcal{S}'_0(\mathbf{R}) \equiv L_2(\mathbf{R}) \to \mathcal{D}'_1(\mathbf{R}) \equiv \mathcal{S}'_1(\mathbf{R})$ is the differential operator observed in § 1.3. (See § 3.4.)

(C). Ornstein–Uhlenbeck equations. A stochastic differential equation of the following type is called an *Ornstein–Uhlenbeck equation* (abbr. *OU-equation*):

$$dX_t = dB_t + AX_t\, dt,$$

where A is a deterministic ($=$ independent of ω) linear operator of $\mathcal{D}'$ (or $\mathcal{D}'_r$)

into $\mathscr{D}'$. The examples mentioned above are *OU*-equations. Because of this particular form we have no difficulty caused by stochastic integrals. This equation is equivalent to

$$\partial_t X_t = \partial_t B_t + A X_t$$

where ∂_t is the derivative in the distribution sense. For each fixed ω this equation is a nonstochastic equation and can be dealt with in the framework of analysis. Nevertheless the solution yields a Markov process and so it is interesting from the probabilistic viewpoint. See [19], [24] and [14].

(D). Stochastic integrals and stochastic differential equations in terms of linear random functionals. In defining stochastic integrals in subsection (A) above, we assumed $(\sigma.1)$ (adaptability) and $(\sigma.2)$ (local integrability) so that the stochastic integrals are $\mathscr{D}'$-processes. Assume the following $(\sigma.2')$ instead of $(\sigma.2)$:

$(\sigma.2')$ (*local integrability in the weak sense*). There exist a sequence of H-seminorms $u_n < \tau(\mathscr{D})$, $n = 1, 2, \ldots$ and an increasing sequence of $\mathscr{F}_.$-stopping times $\theta_n = \theta_n(\omega) \uparrow \infty$ a.s. such that

$$E\left(\int_0^\infty \|\sigma_{s \wedge \theta_n}\|_{u_n}^2 \, ds \right) < \infty,$$

where $\|\cdot\|_u = \|\cdot\|_{p,u}$. By the same procedure as in (A) we can define the stochastic integral

$$I_t(\sigma) = \int_0^t \sigma_s \, dB_s, \qquad t \in T$$

to obtain a family of linear random functionals indexed by $t \in T$. These linear random functionals are continuous with respect to the countably Hilbertian topology τ_u on $\mathscr{D}$ determined by the family $\{u_n\}$. Since τ is nuclear, we can find v_n such that $u_n <_{HS} v_n$. Then the countably Hilbertian topology τ_v determined by $\{v_n\}$ has the property that $\tau_u <_{HS} \tau_v$. Since $I_t(\sigma)$ is a τ_u-continuous linear functional on $\mathscr{D}$, it has a τ_v-regular version. Denoting this version by $I_t(\sigma)$ again, we can regard $I_t(\sigma)$ as a $\mathscr{D}'$-variable. Hence there is no essential difference between this definition and that in (A).

We can formulate the stochastic differential equation (11) or (12) by regarding X_t, $\int_0^t \sigma_s(X_s) \, dB_s$ and $b(X_t)$ as linear random functionals. Solving the equation in this sense and taking regular versions, we will eventually arrive at the same results as in (B).

3.2. Ornstein–Uhlenbeck equations of the Malliavin type.

In this section we will seek the solution of a *Malliavin equation* (an Ornstein–Uhlenbeck equation of the Malliavin type):

$$dX_t = dB_t - \tfrac{1}{2} X_t \, dt, \qquad X_0 = \xi, \tag{1}$$

where $\{B_t\}$ is a Wiener $\mathscr{D}'$-process $(0, p^2)$ and $\{X_t\}$ is a sample continuous

$\mathcal{D}'$-process. This equation is of Type 1, because $\frac{1}{2}X_t$ belongs to $\mathcal{D}'_r$ if X_t does. Writing equation (1) in the integral form, we obtain

$$X_t = \xi + B_t - \frac{1}{2}\int_0^t X_s \, ds,$$

namely

$$X_t(f) = \xi(f) + B_t(f) - \frac{1}{2}\int_0^t X_s(f) \, ds, \qquad f \in \mathcal{D}.$$

Since the topology $\tau = \tau(\mathcal{D})$ is nuclear and since $p \prec \tau$, we can find $q \prec \tau$ such that $p \prec_{HS} q$. The dual space $\mathcal{D}'_q$ with the dual norm q', $(\mathcal{D}'_q, q')$ is a separable Hilbert space. The Wiener $\mathcal{D}'$-process $(0, p)$ is regarded as a sample q'-continuous $\mathcal{D}'_q$-process and so it is a sample continuous $\mathcal{D}'$-process.

THEOREM 3.2.1. *Equation (1) has a unique solution and the solution is a sample q'-continuous centered Gaussian $\mathcal{D}'_q$-process expressible as*

$$X_t(f) = e^{-t/2}\xi(f) + \int_0^t e^{-(t-s)/2} \, dB_s(f), \qquad f \in \mathcal{D}. \tag{2}$$

Proof. For each $f \in \mathcal{D}$, $\{B_t(f)\}_t$ is $p(f)^2$ times the one-dimensional Wiener process and so (1) is a one-dimensional Ornstein–Uhlenbeck equation. Solving this we see that (2) holds a.s. for each pair (t, f). Denote the integral part by $Y_t(f)$. Then $\{Y_t(f), t \in T, f \in \mathcal{D}\}$ is a centered Gaussian system with covariance functional:

$$E(Y_t(f)Y_s(g)) = \alpha(t, s)p(f, g), \tag{3}$$

where

$$\alpha(t, s) = e^{-|t-s|/2} - e^{-(t+s)/2}. \tag{4}$$

Hence

$$\left\|\sum_{i=1}^n c_i Y_t(f_i)\right\|_2^2 = \alpha(t, t)p\left(\sum_{i=1}^n c_i f_i\right)^2,$$

especially

$$\|Y_t(f)\|_2^2 = \alpha(t, t)p(f)^2,$$

where $\|\cdot\|_2$ denotes the norm in $L_2(\Omega)$. These two equalities imply that $Y_t : \mathcal{D} \to L_2(\Omega)$ is linear and p-continuous for each t. Since $L_2(\Omega) \subset L_0(\Omega)$ and $\|\cdot\|_2 \geq \|\cdot\|_0$, Y_t is a p-continuous linear random functional on $\mathcal{D}$ for every t. Since $p \prec_{HS} q$, we can use Theorem 2.3.2 to obtain a q-regular version of Y_t for every t. Denoting the version by the same notation Y_t, we obtain a $\mathcal{D}'_q$-process (and hence a $\mathcal{D}'$-process) $\{Y_t\}$. But (3) implies that

$$E((Y_t - Y_s)(f)^2) = (\alpha(t, t) - 2\alpha(t, s) + \alpha(s, s))p(f)^2$$
$$\leq 2\,|t - s|\,p(f)^2.$$

Since $Y_t - Y_s$ is a centered Gaussian $\mathcal{D}'_q$-variable, we can use Theorem 2.7.3 to check that

$$E(q'(Y_t - Y_s)^4) \leq 3(2\,|t-s|)^2 (p:q)^4_{HS},$$

which implies, by Kolmogorov's continuous version theorem, that $\{Y_t\}$ has a sample q'-continuous version. Denoting the version by $\{Y_t\}$ again and defining

$$X^\xi_t := e^{-t/2}\xi + Y_t, \tag{5}$$

we obtain a sample continuous centered Gaussian $\mathscr{D}'$-process $\{X^\xi_t\}$ satisfying (1). Now it is easy to check that $\{X^\xi_t\}$ is the only one solution of (1). $\square$

If $\xi \in \mathscr{D}'_q$, then $\{X^\xi_t\}$ is a sample q'-continuous $\mathscr{D}'_q$-process.

As in the finite dimensional case we can construct a diffusion process $\{X_t\}$ with state space $\mathscr{D}'$, path space $\mathbf{C} = C(T \to \mathscr{D}')$ and path distributions:

$$P_\xi(\Gamma) = P(X^\xi_\cdot \in \Gamma), \qquad \Gamma \in \mathscr{B}_K(\mathbf{C}), \quad \xi \in \mathscr{D}'.$$

Since $Y_t \in \mathscr{D}'_q$ for every t, we see that if $X^\xi_0 \equiv \xi \in \mathscr{D}'_q$, then X^ξ_t stays in $\mathscr{D}'_q$ for every t by (5). Therefore $\mathscr{D}'_q$ is an invariant part of the state space $\mathscr{D}'$. Similarly every coset of $\mathscr{D}'/\mathscr{D}'_q$ is also an invariant part and the behavior of the diffusion process on any coset can be determined from that on $\mathscr{D}'_q$ by shifting by a deterministic $\mathscr{D}'$-process of the form $e^{-t/2}\eta$ $(\eta \in \mathscr{D}'_q)$. Hence we can restrict our discussion to $\mathscr{D}'_q$, so that the state space, the path space and the path distributions turn out to be $\mathscr{D}'_q$, $\mathbf{C}_q = C(T \to \mathscr{D}'_q)$ and $P_\xi(\Gamma)$, $\Gamma \in \mathscr{B}_K(\mathbf{C}_q)$, $\xi \in \mathscr{D}'_q$ respectively.

Let $\{p_t(\xi, B), B \in \mathscr{B}_K(\mathscr{D}'_q)\}_{t,\xi}$ be the transition probability measures of the diffusion process. Then $p_t(\xi, B) = P(X^\xi_t \in B)$ and so $p_t(\xi, \cdot)$ is a Gauss distribution on $\mathscr{D}'_q$ with

mean functional: $E(X^\xi_t(f)) = e^{-t/2}\xi(f)$,

variance functional: $V(X^\xi_t(f)) = (1 - e^{-t})p(f)^2$,

i.e.

$$p_t(\xi, \cdot) = N(e^{-t/2}\xi, (1 - e^{-t})p^2). \tag{6}$$

But

$$C_{p_t(\xi,\cdot)}(f) = \exp\{ie^{-t/2}\xi(f) - \tfrac{1}{2}(1 - e^{-t})p(f)^2\}$$

$$\xrightarrow[t\uparrow\infty]{} \exp\{-\tfrac{1}{2}p(f)^2\} = C_{N(0,p^2)}(f),$$

and

$$\mu(\cdot) = \int \mu(d\xi)p_t(\xi, \cdot), \qquad t \in T$$

$$\Rightarrow C_\mu(f) = \int \mu(d\xi)C_{p_t(\xi,\cdot)}(f) \xrightarrow[t\uparrow\infty]{} C_{N(0,p^2)}(f)$$

$$\Rightarrow \mu = N(0, p^2).$$

Hence $N(0, p^2)$ is the only invariant measure of this diffusion process.

The transition operator T_t is given by

$$T_t F(\xi) := \int_{\mathscr{D}_q'} p_t(\xi, d\eta) F(\eta) = E(F(X_t^\xi)). \tag{7}$$

Let $\mathbf{L}_\lambda := L_\lambda(\mathscr{D}_q', \mathscr{B}_K(\mathscr{D}_q'), N(0, p^2))$ for every $\lambda \in [1, \infty)$. Then we obtain

$$T_t : \mathbf{L}_\lambda \to \mathbf{L}_\lambda \quad \text{(linear)}, \tag{8}$$

$$\|T_t F\|_\lambda \le \|F\|_\lambda, \qquad F \in \mathbf{L}_\lambda, \tag{9}$$

$$\lim_{t \downarrow 0} \|T_t F - F\|_\lambda = 0, \qquad F \in \mathbf{L}_\lambda. \tag{10}$$

Hence $\{T_t^\lambda := T_t|_{\mathbf{L}_\lambda}\}_t$ is a strongly continuous semigroup of linear contractions on $\mathbf{L}_\lambda$. (8) and (9) follow from the invariance of N. (10) can be verified by noting that the bounded $\mathscr{B}_K(\mathscr{D}_q')$-measurable tame functions are dense in $\mathbf{L}_\lambda$ for every $\lambda \in [1, \infty)$, where a function $F : \mathscr{D}' \to \mathbf{R}$ of the form $F(\xi) = u(\xi(f_1), \xi(f_2), \ldots, \xi(f_n))$ $(u : \mathbf{R}^n \to \mathbf{R})$ is called a *tame function*. Let $\mathscr{L}_\lambda$ be the Hille–Yosida generator of $\{T_t^\lambda\}_t$. If $\lambda \ge \mu$, then

$$\mathbf{L}_\lambda \subset \mathbf{L}_\mu, \qquad \|\xi\|_\lambda \ge \|\xi\|_\mu \qquad (\xi \in \mathbf{L}_\mu),$$

$$\mathscr{D}(\mathscr{L}_\lambda) \subset \mathscr{D}(\mathscr{L}_\mu), \qquad \mathscr{L}_\lambda = \mathscr{L}_\mu |_{\mathscr{D}(\mathscr{L}_\lambda)}.$$

Defining $\mathscr{L} := \overrightarrow{\lim} \mathscr{L}_\lambda$, we obtain

$$\mathscr{D}(\mathscr{L}) = \bigcap_\lambda \mathscr{D}(\mathscr{L}_\lambda), \ F \in \mathscr{D}(\mathscr{L}) \Rightarrow \mathscr{L}F = \mathscr{L}_\lambda F.$$

THEOREM 3.2.2. *If $\{e_k\}$ is an ONB on $(\mathscr{D}, p)$ and if $u \in C^2(\mathbf{R}^n)$ and $u, \partial_i u, \partial_i \partial_j u$ are bounded, then for*

$$F(\xi) := u(\xi(e_1), \xi(e_2), \ldots, \xi(e_n)) \in \mathscr{D}(\mathscr{L})$$

we obtain

$$\mathscr{L}F(\xi) = \frac{1}{2} \sum_{i=1}^n (\partial_i^2 - \xi(e_i) \, \partial_i) u(\xi(e_1), \xi(e_2), \ldots, \xi(e_n)).$$

Hence $\mathscr{D}(\mathscr{L})$ is $\|\cdot\|_\lambda$-dense in $\mathbf{L}_\lambda$ for every λ.

Proof. It suffices to note the following:

$$T_t F(\xi) = E(F(X_t^\xi)) \qquad (X_t = X_t^\xi),$$

$$dB_t(e_i) \, dB_t(e_j) = \delta_{ij} p(e_i, e_j) \, dt = \delta_{ij} \, dt,$$

$$du(X_t^\xi(e_1), X_t^\xi(e_2), \ldots, X_t^\xi(e_n))$$

$$= \sum_i \partial_i u \, dX_t^\xi(e_i) + \tfrac{1}{2} \sum_{ij} \partial_i \partial_j u \, dX_t^\xi(e_i) \, dX_t^\xi(e_j)$$

$$= \sum_i \partial_i u \, dB_t(e_i) + \tfrac{1}{2} \sum_i (\partial_i^2 - X_t^\xi(e_i) \, \partial_i) u \, dt.$$

Example 1. The following special case was observed by P. Malliavin in his theory of stochastic calculus of variation [20], [28], [12]:

$$\mathscr{D} = \mathscr{D}(I), \qquad I = \prod_{i=1}^{d} [0, a_i] \qquad (0 < a_i < \infty),$$

$$p(f)^2 = \int_I \int_I (x \wedge y) f(x) f(y) \, dx \, dy,$$

$$x = (x_1, x_2, \ldots, x_d), \quad y = (y_1, y_2, \ldots, y_d), \quad x \wedge y = \prod_i x_i \wedge y_i.$$

In the general observation above we treated $\{B_t\}$ and $\{X_t^\xi\}$ as a sample q'-continuous $\mathscr{D}_q'$-process and obtained

$$X_t^\xi = e^{-t/2} \xi + Y_t, \qquad Y_t = \int_0^t e^{-(t-s)/2} \, dB_s. \tag{11}$$

For each t, $\{B_t(f)\}_f$ is centered Gaussian and

$$E(B_t(f) B_t(g)) = t p(f, g) = t \int_I \int_I (x \wedge y) f(x) g(y) \, dx \, dy.$$

Hence $\{B_t(f)\}_f$ has a $C(I)$-version $(B_t(x); x \in I)$ for each t which is t times a (d-dimensional) Brownian sheet on I (§ 2.6, Example 2). Then it is easy to see that

$$E(B_t(x) B_s(y)) = (t \wedge s) \cdot (x \wedge y).$$

Since $\{B_t(x)\}_{t,x}$ is a centered Gaussian system, this has a $C(T \times I)$-version $(B_t(x), (t, x) \in T \times I)$, a $((d+1)$-dimensional) Brownian sheet on $T \times I$. Hence $\{B_t\}$ is a sample $\|\cdot\|_\infty$-continuous $C(I)$-process ($\|\cdot\|_\infty$ = maximum norm).

For each t $\{Y_t(f)\}_f$ is centered Gaussian and

$$E(Y_t(f) Y_t(g)) = (1 - e^{-t}) p(f, g) \qquad \text{(see (3))}.$$

Hence Y_t has a $C(I)$-version $(Y_t(x), x \in I)$, which is $(1 - e^{-t})$ times a Brownian sheet. Then

$$Y_t(f) = \int_I Y_t(x) f(x) \, dx,$$

which, combined with (3), implies that

$$\int \int E(Y_t(x) Y_s(y)) f(x) g(y) \, dx \, dy = \int \int \alpha(t, s)(x \wedge y) f(x) g(y) \, dx \, dy,$$

so

$$E(Y_t(x) Y_s(y)) = \alpha(t, s)(x \wedge y). \tag{12}$$

Hence

$$E[(Y_t(x) - Y_s(y))^2] \leq O(\|(t, x) - (s, y)\|) \qquad (\|\cdot\| = \text{norm in } \mathbf{R}^{1+d})$$

holds when t and s move on a bounded subset of T. Since $Y_t(x) - Y_s(y)$ is centered Gaussian,

$$E[(Y_t(x) - Y_s(y))^{2m}] \leqq O(\|(t, x) - (s, y)\|^m).$$

Using Totoki's extension of Kolmogorov's continuous version theorem, we can check that $\{Y_t(x)\}_{t,x}$ has a continuous version. Hence $\{Y_t\}$ is also a sample $\|\cdot\|_\infty$-continuous $C(I)$-process.

Using the continuous versions, we can write (1) as follows:

$$dX_t(x) = dB_t(x) - \tfrac{1}{2}X_t(x)\,dt, \qquad x \in I. \tag{13}$$

This is Malliavin's equation.

If $\xi \in C = C(I)$, then X_t^ξ is also a sample $\|\cdot\|_\infty$-continuous C-process. Similarly to the general case we can define a diffusion process with state space C. The invariant measure is the probability distribution N_0 of the Brownian sheet on I. But the invariant measure N that was defined on $\mathcal{D}_q'$ gives probability 1 to C and $N_0(B) = N(B \cap C)$. Hence there is no essential difference between $L_\lambda(\mathcal{D}_q', N)$ and $L_\lambda(C, N)$ and similarly for $\mathcal{L}$.

Example 2. Let M be a d-dimensional Riemannian manifold with metric tensor g and $\mathcal{D} = \mathcal{D}(m, n)(M)$ the C^∞ fields of tensors of type (m, n) with compact support. Let p be an H-norm$< \tau(\mathcal{D})$ defined by

$$p(f)^2 = \int_M \|f(x)\|^2 \sqrt{g}\,dx \qquad (\sqrt{g}\,dx = \text{volume element})$$

where $\|f\|$ is the g-norm of the tensor f, namely

$$\|f\|^2 = g^{i_1 j_1} \cdots g^{i_n j_n} g_{k_1 l_1} \cdots g_{k_m l_m} f^{k_1 \cdots k_m}_{i_1 \cdots i_n} f^{l_1 \cdots l_m}_{j_1 \cdots j_n}.$$

Let q be an H-norm on $\mathcal{D}$ such that $p <_{HS} q$. Let $\{B_t\}$ be a Wiener $\mathcal{D}'$-process. Then Theorem 2.7.1 ensures that the Ornstein–Uhlenbeck equation (1) has a solution (2), which is a sample q'-continuous Gaussian $\mathcal{D}_q'$-process.

In the special case $\mathcal{D} = \mathcal{D}(0, 0)(I)$, $I = \prod_{i=1}^d (0, a_i)$ the Wiener process $\{B_t\}$ and the solution $\{X_t\}$ in this example are obtained by applying $\partial = \partial_1 \partial_2 \cdots \partial_n$ to those in Example 1, respectively. The invariant measure N here is the image measure of the N in Example 1 by ∂, so the $\mathbf{L}_\lambda$ and $\mathcal{L}$ here are related to those in Example 1 respectively in the obvious way.

Example 3. We can generalize Example 2 as follows: Let M be a d-dimensional manifold (not necessarily Riemannian) and $\mathcal{D} = \mathcal{D}(m, n)$ the C^∞-fields of tensors of type (m, n) with compact support. Let $\nu(dx\,dy)$ be a locally finite measure on $M \times M$ symmetric in (dx, dy). Let

$$\theta(x, y) = \theta^{i_1, \ldots, i_n, j_1, \ldots, j_n}_{k_1, \ldots, k_m, l_1, \ldots, l_m}(x, y)$$

be a locally bounded and Borel measurable tensor field of type $(2n, 2m)$ on the product manifold $M \times M$ that is symmetric and positive definite in $(i_1, \ldots, i_n, k_1, \ldots, k_m, x)$ and $(j_1, \ldots, j_n, l_1, \ldots, l_m, y)$. Define $p(f)$ to be the square root of

$$\int_M \int_M \theta^{i_1, \ldots, i_n, j_1, \ldots, j_n}_{k_1, \ldots, k_m, l_1, \ldots, l_m}(x, y) f^{k_1 \cdots k_m}_{i_1 \cdots i_n}(x) f^{l_1 \cdots l_m}_{j_1 \cdots j_n}(y) \nu(dx\,dy).$$

Then p is a Hilbertian seminorm on $\mathscr{D}$ such that $p < \tau(\mathscr{D})$. Now we can define a Wiener $\mathscr{D}'$-process $(0, p^2)$, $\{B_t\}$ and then an Ornstein–Uhlenbeck process $\{X_t\}$ by virtue of Theorem 3.2.1. In Example 2 ν is concentrated on the diagonal set of $M \times M$.

If (i) M is Riemannian, (ii) $\nu(dx\,dy) = \sqrt{g(x)g(y)}\,dx\,dy$ ($\sqrt{g(x)}\,dx = $ volume element in M) and $\theta(x, y)$ is C^∞ in (x, y), we can observe the random $\mathscr{D}'$-variables $\{B_t(f)\}_f$ in each chart defining the structure of M and use an argument similar to that used in Example 4 of § 2.6 to conclude that almost every sample functional of B_t is a C^∞-function on M for each t.

3.3. Properties of $\mathscr{L}_2$. In the last section we introduced an operator $\mathscr{L}_2$ in $\mathbf{L}_2 = L_2(\mathscr{D}'_q, \mathscr{B}_K(\mathscr{D}'_q), N)$. Since $\mathscr{L}_2$ is the generator of the strongly continuous semigroup $\{T_t^2\}_t$ of linear contractions on $\mathbf{L}_2$, $\mathscr{L}_2$ is a closed linear operator densely defined on $\mathbf{L}_2$.

Let $\{e_n\}$ be an ONB in the separable Hilbert space $(\mathscr{D}_q, q)$. Since N is a Gauss measure on $(\mathscr{D}'_q, \mathscr{B}_K(\mathscr{D}'_q))$ with mean functional 0 and variance functional p^2 ($p <_{HS} q$),

$$\{\xi(e_n), n = 1, 2, \ldots\} \qquad (\xi \in \mathscr{D}'_q) \tag{1}$$

is a family of real random variables that are independently identically $N_{0,1}$-distributed on the probability space $= (\mathscr{D}'_q, \mathscr{B}_K(\mathscr{D}'_q), N)$. Every element of $\mathbf{L}_2$ can be approximated by tame functions of the form $F(\xi) = u(\xi(e_1), \ldots, \xi(e_n))$, $u \in \mathscr{D}(\mathbf{R}^n)$. Let $\mathscr{F}$ denote the space of all such tame functions. Then $\mathscr{F}$ is a dense vector subspace of $\mathbf{L}_2$ and

$$\mathscr{L}_2 F(\xi) = \tfrac{1}{2} \sum_{i=1}^{n} (\partial_i^2 - \xi(e_i)\,\partial_i)u(\xi(e_1), \ldots, \xi(e_n)) \tag{2}$$

for $E(\xi) = u(\xi(e_1), \ldots, \xi(e_n))$. If $F \in \mathscr{F}$, then $\mathscr{L}_2 F \in \mathscr{F}$. If $F_i = u_i(\xi(e_1), \ldots, \xi(e_n)) \in \mathscr{F}$, then

$$(\mathscr{L}_2 F_1, F_2)_{\mathbf{L}_2}$$

$$= \int_{\mathbf{R}^n} \frac{1}{2} \sum_{i=1}^{n} (\partial_i^2 - x_i\,\partial_i)u_1(x_1, \ldots, x_n) \cdot u_2(x_1, \ldots, x_n) \prod_{i=1}^{n} N_{0,1}(dx_i)$$

$$= -\frac{1}{2} \int_{\mathbf{R}^n} \sum_i \partial_i u_1\,\partial_i u_2 \prod_{i=1}^{n} N_{0,1}(dx_i) \quad \text{(by integration by parts)}.$$

Hence

$$(\mathscr{L}_2 F_1, F_2)_{\mathbf{L}_2} = (F_1, \mathscr{L}_2 F_2)_{\mathbf{L}_2}, \tag{3}$$

so the restriction $\mathscr{L}_2|_{\mathscr{F}}$ is linear and symmetric.

For every sequence $\mathbf{n} = (n_1, n_2, \ldots)$ ($n_i = 0, 1, 2, \ldots$) such that $n_i = 0$ except for finitely many i's, we set

$$\mathbf{H_n}(\xi) = \prod_i H_{n_i}(\xi(e_i)/\sqrt{2}), \qquad H_n : \text{Hermite} \tag{4}$$

$$(H_n = \text{Hermite polynomial of degree } n).$$

Note that the factors equal 1 except for finitely many i's by $H_0 = 1$. Since $\{H_n(x/\sqrt{2})\}_n$ is an orthogonal base in $L_2(\mathbf{R}, N_{0,1})$ and since $\mathcal{F}$ is dense in $\mathbf{L}_2$, we can easily see that $\{\mathbf{H_n}(\xi)\}_\mathbf{n}$ is an orthogonal base in $\mathbf{L}_2$. Since

$$(\partial^2 - x\,\partial)H_n(x/\sqrt{2}) = -nH_n(x/\sqrt{2}),$$

equality (2) implies that

$$\mathcal{L}_2\mathbf{H_n} = -\tfrac{1}{2}\,|\mathbf{n}|\,\mathbf{H_n}\left(|(\mathbf{n})| = \sum_i n_i\right). \tag{5}$$

Since $\mathcal{L}_2$ is a closed linear operator, $\mathcal{L}_2$ is self-adjoint with eigenvalues $\{0, -\tfrac{1}{2}, -1, \ldots\}$ and eigenvectors $\{\mathbf{H_n}\}$. Let $\mathcal{H}_n$ denote the closed linear span of

$$\mathbf{H_n}, \qquad |\mathbf{n}| = n.$$

Then $\mathcal{H}_n$ is the eigenspace of $\mathcal{L}_2$ corresponding to $-n/2$. Thus we have

$$\mathbf{L}_2 = \sum_n \oplus\, \mathcal{H}_n, \tag{6}$$

$$\mathcal{L}_2 = \sum_n -\frac{n}{2}\,\not{p}_n \qquad (\not{p}_n = \text{projection to } \mathbf{H}_n). \tag{7}$$

Example. Let $\mathcal{D} = \mathcal{D}(\mathbf{R}^d)$ and

$$p(f) = \|f\|_{L_2(\mathbf{R}^d)}. \tag{8}$$

Then N is the probability distribution of the white noise on $\mathbf{R}^d$. Hence for every $\xi \in \mathcal{D}_q'$, $(p <_{HS} q)$ the linear random functional $\xi : \mathcal{D} \to \mathbf{L}_2 \equiv L_2(\mathcal{D}_q', N)$ can be extended to a linear random functional $\bar{\xi} : L_2(\mathbf{R}^d) \to \mathbf{L}_2$. Hence

$$M(A) = \bar{\xi}(1_A), \qquad |A| < \infty \quad (|\cdot| = \text{Lebesgue measure})$$

defines a Gaussian random measure on $\mathbf{R}^d$ on the probability space $(\mathcal{D}_q', \mathcal{B}_K(\mathcal{D}_q'), N)$. It is obvious that

$$(M(A), 1)_{\mathbf{L}_2} = 0, \qquad (M(A_1), M(A_2))_{\mathbf{L}_2} = |A_1 \cap A_2|.$$

We can define the *multiple Wiener integral* [15], [13]:

$$I_n(f_n) = (n!)^{-1/2}\underset{(\mathbf{R}^d)^n}{\int \cdots \int} f_n(x_1, x_2, \ldots, x_n)M(dx_1)M(dx_2)\cdots M(dx_n).$$

Let $\tilde{L}_2^n(\mathbf{R}^d)$ be the Hilbert space of all symmetric square integrable functions on $\mathbf{R}^d$. Then $I_n : \tilde{L}_2^n(\mathbf{R}^d) \to \mathcal{H}_n$ is isomorphic. $\tilde{L}_2^n(\mathbf{R}^d)$ is the symmetric tensor product of $\mathcal{H} := L_2(\mathbf{R}^d)$ and so is denoted by $S_n\mathcal{H}^n$. (6) can be written as

$$\mathbf{L}_2 \cong \sum_n \oplus\, S_n\mathcal{H}^n. \tag{9}$$

This shows that $\mathbf{L}_2$ is a representation of the symmetric (or Boson) *Fock space* [25] where $-2\mathcal{L}_2$ is called the *number operator* in view of (7).

3.4. Ornstein–Uhlenbeck processes of the Gaveau type. Let $\{B_t\}$ be a Wiener $\mathscr{D}'$-process $(0, p^2)$ and K a strictly positive definite self-adjoint HS-operator on the Hilbert space $(\mathscr{D}'_p, p')$. We want to find a $\mathscr{D}'$-process X_t satisfying the stochastic differential equation:

$$dX_t = dB_t - \tfrac{1}{2}K^{-2}X_t \, dt, \qquad X_0 = \xi. \tag{1}$$

This equation is called an *Ornstein–Uhlenbeck equation of the Gaveau type*. It is not of Type 1, because $K^{-2}X_t \in \mathscr{D}'_r$ even if $X_t \in \mathscr{D}'_r$.

We will solve this under the restriction that X_t is a $\mathscr{D}'_p$-process, so $\xi = X_0 \in \mathscr{D}'_p$. By the assumption on K we have an ONB $\{\varepsilon_n\}$ on $(\mathscr{D}'_p, p')$ such that

$$K\varepsilon_n = \lambda_n \varepsilon_n, \qquad \lambda_n > 0, \quad \sum \lambda_n^2 < \infty.$$

Let $\{e_n\}$ be the dual ONB of $\{\varepsilon_n\}$ on $(\bar{\mathscr{D}}_p, \bar{p})$. Since $(\mathscr{D}'_p, p')$ is also the dual space of $(\bar{\mathscr{D}}_p, \bar{p})$ and since X_t is a $\mathscr{D}'_p$-process, we can define $X_t(f)$, $f \in \bar{\mathscr{D}}_p$ for every ω. Since $\{B_t(f), f \in \mathscr{D}\}$ is a linear random functional on $\mathscr{D}$ with $E(B_t(f)^2) = tp(f)^2$, it is regarded as a linear random functional on $\bar{\mathscr{D}}_p$, namely $B_t(f)$ is a real random variable for every $f \in \bar{\mathscr{D}}_p$, although the sample functional B_t does not necessarily belong to $\mathscr{D}'_p$; in fact it belongs to $\mathscr{D}'_q$ for any q such that $p <_{HS} q$.

Hence, for each $f \in \bar{\mathscr{D}}_p$ we have

$$dX_t(f) = dB_t(f) - \tfrac{1}{2}K^{-2}X_t(f) \, dt \quad \text{a.s.}$$

Setting $f = e_n$ and noting that

$$K^{-2}X_t(e_n) = p'(K^{-2}X_t, \varepsilon_n) = p'(X_t, K^{-2}\varepsilon_n)$$
$$= \lambda_n^{-2}p'(X_t, \varepsilon_n) = \lambda_n^{-2}X_t(e_n),$$

we obtain a countable number of equations:

$$dX_t(e_n) = dB_t(e_n) - \tfrac{1}{2}\lambda_n^{-2}X_t(e_n) \, dt, \qquad n = 1, 2, \ldots \tag{2}$$

holding simultaneously a.s. Since $\{B_t(e_n)\}_t$ is a sample-continuous centered Gaussian process with

$$E(B_t(e_n)B_s(e_n)) = (t \wedge s)\bar{p}(e_n, e_n) = (t \wedge s), \tag{3}$$

$\{B_t(e_n)\}_t$ is a one-dimensional Wiener process for every n. Since $\{B_t(e_n)\}_{t,n}$ is a centered Gaussian system with

$$E(B_t(e_m)B_s(e_n)) = (t \wedge s)p(e_m, e_n) = 0 \qquad (m \neq n), \tag{4}$$

$\{B_\cdot^{(n)} = B_\cdot(e_n)\}_{n=1,2,\ldots}$ is an independent sequence of one-dimensional Wiener processes.

For each n equation (2) is a one-dimensional Ornstein–Uhlenbeck equation, so the solution is

$$X_t^{(n)} = X_t(e_n) = \xi^{(n)}e^{-\lambda_n^{-2}t/2} + \int_0^t e^{-\lambda_n^{-2}(t-s)/2} \, dB_s^{(n)}, \tag{5}$$

where $\xi^{(n)} = \xi(e_n)$. $X_t^{(n)}$ is an Ornstein–Uhlenbeck process and so it is sample continuous, and

$$E(X_t^{(n)}) = \xi^{(n)} e^{-\lambda_n^{-2}t/2},$$
$$V(X_t^{(n)}, X_s^{(n)}) = \lambda_n^2(e^{-\lambda_n^{-2}|t-s|/2} - e^{-\lambda_n^{-2}(t+s)/2}),$$
$$V(X_t^{(n)}) = \lambda_n^2(1 - e^{-\lambda_n^{-2}t}), \quad V(X_t^{(m)}, X_s^{(n)}) = 0 \quad (m \neq n).$$

Denote the two terms of the right-hand side of (5) by $\xi_t^{(n)}$ and $Y_t^{(n)}$ and define

$$\xi_t := \sum_n \xi_t^{(n)} \varepsilon_n, \qquad Y_t := \sum_n Y_t^{(n)} \varepsilon_n. \tag{6}$$

Since $|\xi_t^{(n)}| \leq |\xi^{(n)}|$ and $E((Y_t^{(n)})^2) = V(X_t^{(n)}) \leq \lambda_n^2$, we obtain

$$\sum_n (\xi_t^{(n)})^2 \leq \sum_n (\xi^{(n)})^2 = p'(\xi)^2 < \infty, \tag{7}$$

$$E\left(\sum_n (Y_t^{(n)})^2\right) \leq \sum \lambda_n^2 < \infty, \tag{8}$$

and so

$$\sum_n (Y_t^{(n)})^2 < \infty \quad \text{a.s.} \tag{9}$$

Hence the infinite sequences of (6) are p'-convergent, so

$$\xi_t \in \mathcal{D}_p' \quad \text{and} \quad Y_t \in \mathcal{D}_p' \quad \text{a.s. for each } t. \tag{10}$$

It is easy to see that ξ_t is p'-continuous in t.

Take $q < \tau(\mathcal{D})$ such that $p <_{HS} q$ and an ONB $\{d_n\}$ on $(\mathcal{D}, q)$. By the definition of $Y_t^{(n)}$ we obtain

$$E((Y_t^{(n)} - Y_s^{(n)})^2) = \lambda_n^2(2 - e^{-\lambda_n^{-2}t} - e^{-\lambda_n^{-2}s} - 2e^{-\lambda_n^{-2}|t-s|/2} + 2e^{-\lambda_n^{-2}(t+s)/2})$$
$$\leq \lambda_n^2(2 - 2e^{-\lambda_n^{-2}|t-s|/2}) \leq |t-s|.$$

Since $\{e_n\}$ is an ONB on $(\mathcal{D}, p)$, we obtain

$$(Y_t - Y_s)(d_n) = \sum_m (Y_t^{(m)} - Y_s^{(m)})p(d_n, e_m),$$

where the right-hand side is an orthogonal series in $L_2(\Omega)$. Hence

$$E((Y_t - Y_s)(d_n)^2 = \sum_m E((Y_t^{(m)} - Y_s^{(m)})^2)p(d_n, e_m)^2$$
$$\leq |t-s| \, p(d_n)^2,$$

$$E(q'(Y_t - Y_s)^2) = E\left(\sum_n (Y_t - Y_s)(d_n)^2\right)$$
$$\leq c \, |t-s|, \qquad c = \sum_n p(d_n)^2 = (p:q)_{HS}^2 < \infty.$$

Since $Y_t - Y_s$ is a centered Gaussian $\mathcal{D}_q'$-variable by $\mathcal{D}_q' \supset \mathcal{D}_p'$, we can use Theorem 2.7.3 to obtain

$$E(q'(Y_t - Y_s)^4) \leq c_1 |t-s|^2 \qquad (c_1 \text{ a constant}).$$

Hence we can use Kolmogorov's theorem to conclude that $\{Y_t\}$ has a continuous version, which we denote by the same notation $\{Y_t\}$.

Now it is easy to check that $\{X_t := \xi_t + Y_t\}$ is a sample continuous $\mathscr{D}'_q$-process (and so a $\mathscr{D}'$-process) satisfying (1).

As in §2, $\{X_t\}$ yields a Markov process, with path space $\mathscr{D}'$ (or $\mathscr{D}'_q$) such that $P(X_t \in \mathscr{D}'_p) = 1$ for every t. Since we are concerned with transition probabilities and transition operators, we can take $\mathscr{D}'_p$ (instead of $\mathscr{D}'$ or $\mathscr{D}'_q$) for the state space. Representing the separable Hilbert space $(\mathscr{D}'_p, p')$ by

$$l_2 = \left\{ \xi = (x_0, x_1, \ldots) : \sum_n x_n^2 < \infty \right\}$$

and observing that $\{X_t^{(n)}\}_n$ is independent, we see that the transition probability is

$$p(t, \xi, \cdot) = \prod_n N(e^{-\lambda_n^{-2}t/2} x_n, \lambda_n^2(1 - e^{-\lambda_n^{-2}t})).$$

THEOREM 3.4.1 (B. Gaveau [8]). *The invariant measure of the Markov process is*

$$\mu = \prod_n N(0, \lambda_n^2),$$

$p(t, \xi, \cdot)$ *is absolutely continuous with respect to* μ, *and the density is the infinite product of the densities of the corresponding factors.*

Proof. Since

$$\int_{l_2} p(t, \xi, d\eta) f(\eta) \to \int_{l_2} \mu(d\eta) f(\eta) \qquad (t \uparrow \infty),$$

for every continuous bounded tame function f, μ is the unique invariant measure. To prove the absolute continuity use the examples of Kakutani's theorem presented in Kuo's note [18, p. 117]. $\quad\square$

The transition operators

$$T_t f(\xi) = \int p_t(\xi, d\eta) f(\eta)$$

form a strongly continuous semigroup of linear contractions on $L_\lambda(l_2, \mu)$ $(\lambda \in [1, \infty))$ and its generator is the closed extension of

$$A = \frac{1}{2} \sum_{n=1}^\infty \left(\frac{d^2}{dx_n^2} - \lambda_n^{-2} x_n \frac{d}{dx_n} \right),$$

$\mathscr{D}(A) = $ the bounded C^2 tame functions.

Example 1. $\mathscr{D}' = \mathscr{D}'(\mathbf{R}^d)$, $p = $ the norm in $L_2(\mathbf{R}^d)$, $K = (\|x\|^2 - \Delta)^{-1}$ applied in the distribution sense,

$$\{e_n\} = \{h_{n_1}(x_1) h_{n_2}(x_2) \cdots h_{n_d}(x_d)\}_{d, n_i}$$

where $\{h_n\}$ is a sequence of functions mentioned in §1.3.

Example 2. (*B. Gaveau* [8]). $\mathscr{D}' = \mathscr{D}(\mathbf{T}^d)$ ($\mathbf{T}^d$ = the *n*-dimensional torus), $K = (-\Delta)^{-1/2}$, p = the norm in $L_2(\mathbf{T}^d)$.

3.5. Funaki's random motion of strings. No difficulty caused by stochastic integrals appears in Ornstein–Uhlenbeck equations. As an example of proper stochastic differential equations we will mention Funaki's equation of a random motion of strings.

As a limit of a sequence of approximate discrete models T. Funaki [7] formally derived a stochastic differential equation of the following type:

$$dY_t(x) = \sigma(Y_t(x))\, dB_t(x) + \left(a(Y_t(x)) + \frac{k}{2}\partial_x^2 Y_t(x)\right) dt \qquad (1)$$

and proved an interesting fact about the asymptotic behaviour of $Y_t(x)$ as the parameter k tends to ∞. Although he discussed the case where $Y_t(x)$ and $B_t(x)$ are $\mathbf{R}^d$-valued, we assume $d = 1$ for simplicity.

What he did was to prove the existence of the limit in distributions of the solutions of approximating stochastic differential equations using Prokhorov's theorem and to investigate the property of the limit process. He called the limit process the solution of the infinite dimensional stochastic differential equation (1).

What we want to do is first to define (1) from our standpoint and next to prove the existence and uniqueness of its solution; this turns out to be the limit process obtained by Funaki. Here we will give only the outline of our idea.

We impose the following assumptions due to Funaki.

(A.1) $\{B_t\}$ is a Wiener $\mathscr{D}'(I)$-process ($I = [0, 1]$) such that B_t is a white noise on I with intensity t.

(A.2) $\{Y_t\}$ is a $\|\cdot\|_\infty$-continuous $C(I)$-process with some boundary conditions, for example $Y_t(0) = Y_t(1) = 0$, and $\{Y_t\}$ is adapted to $\{B_t\}$.

(A.3) $\sigma : \mathbf{R} \to [0, \infty)$ and $a : \mathbf{R} \to \mathbf{R}$ satisfies Lipschitz conditions.

Since $B_t(x)$ and $\partial_x^2 Y_t(x)$ are not functions but Schwartz distributions in x, we should interpret (1) as a stochastic differential equation in $\mathscr{D}'(I)$ (§ 1.4).

$\{B_t\}$ is a Wiener $\mathscr{D}'(I)$-process $(0, p^2)$ where

$$\mathscr{D} = \mathscr{D}(I), \qquad p = \|\cdot\| \quad (\text{norm in } L_2(I)).$$

Since $Y_t(x)$ is continuous in $x \in I$, Y_t is a $\mathscr{D}_0'(I)$-variable. But $\partial^2 Y_t$ is a $\mathscr{D}_1'$-variable. Hence (1) is of Type 2 (§ 3.1(B)).

Since $E(B_t(f)^2) = t\,\|f\|^2$, $B_t \in \mathscr{D}_{1/2}'(I)$ ($\subset \mathscr{D}_1'(I)$) by the regularization theorem (§ 2.5). If σ is bounded, the multiplication operator acting on $B_t - B_s$ belongs to $\mathscr{L}_{0,1/2}$. Hence the stochastic integral

$$I_t(\sigma) = \int_0^t \sigma(Y_s(\cdot))\, dB_s$$

is well defined. If σ is not bounded, we define it using a sequence of stopping times

$$\theta_n = \inf \left\{ t : \max_{x \in I} |\sigma(Y_t(x))| > n \right\}, \qquad n = 1, 2, \ldots.$$

Thus the meaning of (1) from our standpoint has become clear.

To solve (1) we use the semigroup of operators $\{U_t, t \geq 0\}$ with generator

$$AU(x) = \frac{k}{2} \frac{d^2}{dx^2} U(x), \qquad U(0) = U(1) = 0$$

to transform (1) into the following stochastic integral equation:

$$Y_t = U_t \xi + \int_0^t U_{t-s}(\sigma(Y_s(\cdot)) \, dB_s) + \int_0^t U_{t-s} a(Y_s(\cdot)) \, ds, \qquad \xi = Y_0. \qquad (2)$$

Note that the stochastic integral above is well defined because U_{t-s} is deterministic.

Viewing Y_t as a process of linear random functionals (§ 3.1(D)), we can solve (2) by translating Dawson's method [3] into our language. Using Totoki's extension of the Kolmogorov continuous version theorem (Theorem 2.6.5) we can prove that the process of linear random functionals $\{Y_t(x)\}$ obtained above has a version continuous in (t, x); see Funaki [7]. Thus we obtain the solution of (1), if σ is bounded. To deal with the unbounded case, we can use the stopping time arguments as in the definition of stochastic integrals.

References

[1] Ja. I. Belopol'skaya and Ja. D. Daleckiĭ, *Diffusion processes in smooth Banach spaces and manifolds* I, Trans. Moscow Math. Soc., 37 (1978); Transl. AMS 1 (1980), pp. 113–150.

[2] P. Billingsley, *Convergence of Probability Measures*, John Wiley, New York, 1968.

[3] D. A. Dawson, *Stochastic evolution equations and related measure processes*, J. Multivariate Anal., 5 (1975), pp. 1–55.

[4] K. D. Elworthy, *Stochastic Differential Equations on Manifolds*, Cambridge Univ. Press, Cambridge, 1982.

[5] X. Fernique, *Processus linéaires et processus généralisés*, Ann. Inst. Fourier, 17 (1967), pp. 1–92.

[6] A. Friedman, *Stochastic Differential Equations and Applications* I, II, Academic Press, New York, 1975.

[7] T. Funaki, *Random motion of strings and related stochastic evolution equations*, Nagoya Math. J., 89 (1983), pp. 129–193.

[8] B. Gaveau, *Noyeau des probabilités de transition de certains opérateurs d'Ornstein–Uhlenbeck dans l'éspace de Hilbert*, C.R. Acad. Sci. Paris, I-293 (1981), pp. 469–472.

[9] I. Gelfand and N. Ya. Vilenkin, *Generalized Functions* 4, Springer, Berlin, 1964.

[10] I. I. Gikhman and A. V. Skorokhod, *Stochastic Differential Equations*, Ergeb. d. Math. 72, Springer, Berlin, 1972; *The Theory of Stochastic Processes* III, Grundl. d. Math. Wiss., 232, Springer, Berlin, 1979.

[11] H. Hoffmann-Jørgensen, *The theory of analytic spaces*, Various Publication Series No. 10, Aarhus University, Aarhus, Denmark, 1970.

[12] N. Ikeda and S. Watanabe, *Stochastic Differential Equations and Diffusion Processes*, Kodansha, Tokyo; North-Holland, Amsterdam, 1981.

[13] K. Itô, *Complex multiple Wiener integrals*, Jap. J. Math., 22 (1952), pp. 63–86.

[14] K. Itô, *Infinite dimensional Ornstein–Uhlenbeck processes*, Proc. Taniguchi International Conference on Stochastic Analysis, 1982, Katata-Kyoto, Kinekuniya, Tokyo, 1984.

[15] K. Itô, *Multiple Wiener integrals*, J. Math. Soc. Japan, 3 (1951), pp. 159–169.

[16] K. Itô and M. Nawata, *Regularization of linear random functionals*, Probability Theory and Mathematical Statistics, Proc. 4th USSR-Japan Symposium, Lecture Notes in Mathematics 1021, Springer, Berlin, 1983.

[17] A. N. Kolmogorov, *A note on the papers of R. A. Minlos and V. Sazonov*, Theory Prob. Appl., 4 (1959), pp. 221–223.

[18] H. H. Kuo, *Gaussian Measures in Banach Spaces*, Lecture Notes in Mathematics 463, Springer, Berlin, 1975.

[19] J. T. Lewis and L. C. Thomas, *A characterization of regular solutions of a linear stochastic differential equation*, Z. Wahrsch. Verw. Gebiete, 30 (1974), pp. 45–55.

[20] P. Malliavin, *Stochastic calculus of variation and hypoelliptic operators*, Proc. International Symposium on Stochastic Differential Equations, Kyoto 1976, Kinokuniya, Tokyo, 1978.

[21] H. P. McKean, *Stochastic Integrals*, Academic Press, New York, 1969.

[22] M. Metivier and J. Pellaumail, *Stochastic Integration*, Academic Press, New York, 1980.

[23] R. A. Minlos, *Generalized random processes and their extension in measure*, Selected Translations in Mathematical Statistics and Probability, No. 3, American Mathematical Society, Providence, RI, 1963, pp. 291–313.

[24] Y. Okabe, *On a stationary Gaussian process with T-positivity and its associated Langevin equation and S-matrix*, J. Fac. Science, Univ. Tokyo, IA, 26-1 (1979), pp. 115–165.

[25] M. Reed and B. Simon, *Methods of Modern Mathematical Physics*, Academic Press, New York, 1972.

[26] V. Sazonov, *On characteristic functionals*, Theory. Prob. Appl., 3 (1958), pp. 188–192.

[27] L. Schwartz, *Théorie des distributions*, Hermann, Paris, 1966, Chapter III.

[28] D. W. STROOCK, *The Malliavin calculus and its application to second order partial differential equations*, Math. Systems Theory, 14 (1981), pp. 25–63, 141–171.

[29] D. W. STROOCK AND S. R. S. VARADHAN, *Multidimensional Diffusion Processes*, Grundl. d. Math. Wiss., 233, Springer, Berlin, 1979.

[30] F. TREVES, *Topological Vector Spaces, Distributions and Kernels*, Academic Press, New York, 1967.

[31] Y. YAMAZAKI, *Measures on Infinite-Dimensional Spaces*, I, II, Kinokuniya, Tokyo, 1978. (In Japanese.)

[32] K. YOSIDA, *Functional Analysis*, Springer, Berlin, 1965.